# Understanding Quality Assurance in Construction

# Also available from E & FN Spon

**Facilities Management**
K Alexander

**The Construction Net**
A Bridges

**Understanding JCT Standard Building Contracts**
D M Chappell

**Construction Quality and Quality Standards**
G A Atkinson

**Creating the Built Environment**
L Holes

**Value Management in Design and Construction**
J Kelly and S Male

**An Introduction to Building Procurement Systems**
J W E Masterman

**Construction Contracts**
J Murdoch and W Hughes

**Building International Construction Alliances**
R Pietroforte

**Understanding the Building Regulations**
S Polley

**Risk Analysis in Project Management**
J Raferty

**Programme Management Demystified**
G Reiss

**Project Management Demystified**
G Reiss

**Risk Avoidance for the Building Team**
B Sawczuk

*For more information about these and other titles please contact:*
The Marketing Department, E & FN Spon, 11 New Fetter Lane, London, EC4P 4EE. Tel: 0171 842 2252
www.efnspon.com

# Understanding Quality Assurance in Construction

## A practical guide to ISO 9000 for contractors

Edited by
**H. W. Chung**
*University of Technology Sydney*

First published 1999 by E & FN Spon
11 New Fetter Lane, London EC4P 4EE

Simultaneously published in the USA and Canada
by Routledge
29 West 35th Street, New York, NY 10001

*E & FN Spon is an imprint of the Taylor & Francis Group*

© 1999 H. W. Chung

Printed and bound in Great Britain by MPG Books Ltd, Bodmin

All rights reserved. No part of this book may be reprinted or reproduced or utilised in any form or by any electronic, mechanical, or other means, now known or hereafter invented, including photocopying and recording, or in any information storage or retrieval system, without permission in writing from the publishers.

The publisher makes no representation, express or implied, with regard to the accuracy of the information contained in this book and cannot accept any legal responsibility or liability for any errors or omissions that may be made.

*British Library Cataloguing in Publication Data*
A catalogue record for this book is available from the British Library

*Library of Congress Cataloging in Publication Data*
Chung, H. W. (Hung W.)
    Understanding quality assurance in construction : a practical guide to ISO9000 / H.W. Chung
        p.   cm.
    Includes bibliographical references and index.
    1. Building --Quality control--Standards.  2. ISO 9000 Series Standards.  I. Title.
TH4382. C48  1999
690'. 0685--dc21                                                 99-20876
                                                                                  CIP

ISBN 0-419-24950-8

*To my wife Juliana*

# Contents

Preface ix

Acknowledgements x

PART ONE   ESTABLISHING A QUALITY SYSTEM

**1   Quality and quality assurance**
1.1  What is quality? 3
1.2  Quality control 4
1.3  Quality assurance 5
1.4  Is quality assurance for construction? 7
1.5  Summary 10

**2   Quality system and system requirements**
2.1  Quality system 11
2.2  Quality system standard 14
2.3  ISO 9000 family of standards 15
2.4  Quality system requirements 18
2.5  Summary 37

**3   Project Quality Management**
3.1  Project quality planning 42
3.2  Quality plan 45
3.3  Inspection and test plan 49
3.4  Project management review 50
3.5  Client's role in quality assurance 52
3.6  Summary 54

**4   Developing a quality system**
4.1  Where to start 57
4.2  Getting started 59
4.3  Reviewing current practices 60
4.4  Preparing quality system documents 63
4.5  Quality related training 65
4.6  Summary 66

## 5 Implementing a quality system
5.1 The trial period — 67
5.2 Motivation — 68
5.3 Training — 70
5.4 Internal quality audits — 71
5.5 Management review — 81
5.6 Summary — 82

## 6 Third party certification
6.1 Quality system audits — 84
6.2 Why third party certification? — 85
6.3 Selecting a certification body — 85
6.4 Scope of certification — 87
6.5 Process of certification — 88
6.6 Preparing for system audit — 94
6.7 Surveillance audit — 95
6.8 Time and cost of certification — 96
6.9 Summary — 97

## 7 Facts and fallacies
7.1 Perceived outcomes of quality system — 99
7.2 Criticisms of ISO 9000 — 100
7.3 ISO 9000 in small businesses — 105
7.4 The way ahead — 107
7.5 Summary — 108

## PART TWO   DOCUMENTING A QUALITY SYSTEM

## 8 Writing quality system documents
8.1 Document layout and format — 113
8.2 Writing the quality manual — 114
8.3 Writing the quality procedures — 118
8.4 Summary — 120

## 9 Sample quality system documents
9.1 Overview — 122
9.2 Quality manual — 125
9.3 Quality procedures — 151

**References** — 247

**Index** — 249

# Preface

Quality is a buzzword of the day. Quality assurance has evolved from a manufacture-centred discipline to one with broad management implications across all industries and professions. The ISO 9000 family of standards has been adopted worldwide as a framework for relationship between the supplier (or service provider) and the customer. There is evidence to show that an organization operating a quality system conforming to ISO 9000 enjoys a competitive edge in its business.

The construction industry has been slow, and somewhat reluctant, to embrace the ISO 9000 concepts of quality assurance in its practice. The phenomenon is particularly apparent in small to medium-sized contracting companies. However, there is a movement especially in the public sector towards employing only quality endorsed contractors and subcontractors for new projects. In the foreseeable future, all contracting companies irrespective of size will be attracted to the benefits, and certainly the business necessity, of following the trend already set by the big names in the industry.

With this book, I hope to offer assistance to those contractors and subcontractors who are at the crossroads. They may have the desire to go for ISO 9000 but do not know how to get there. The apparently daunting task is made easy with a step-by-step approach. The sequence of actions involved is simply explained and illustrated with examples. Included in the book is a quality manual, together with a set of quality procedures, for a hypothetical (but typical) building construction company. The quality system described is geared to the conventional practice and organization structure of a medium-sized company. Although it is not intended to be taken off the shelf and applied right away, the documentation may be used as a template for writing up the quality system of your own.

While every effort is made to ensure accuracy and completeness of information, no warranty is expressed or implied as to the opinions and documentation contained in the book. The material presented is for guidance only and carries no legal or professional liability.

H. W. Chung  
March 1999, Sydney

# Acknowledgements

In the course of preparation of the book, I interviewed a number of prominent professionals in building construction in Australia, Singapore and Hong Kong. They were kind enough to share with me their experiences of applying ISO 9000 that had been accumulated over the years. To them I wish to express my sincere gratitude. I am especially grateful to Mr. Joseph Shek, Managing Director, Excel Engineering Company Limited, Hong Kong, for arranging some of the interviews and for constructive comments on the manuscript. Thanks are also due to Associate Professor Marton Marosszeky, Director, Australian Centre for Construction Innovation, University of New South Wales, for the insight of the Australian construction practice that I have gained through numerous discussions.

The book contains excerpts and illustrations taken from publications of the Construction Industry Research and Information Association (CIRIA), Standards Australia / Standards New Zealand and Master Builders Australia. Permission for reproduction granted by these organizations is gratefully acknowledged. For further information and explanation, readers are referred to the publications themselves.

In addition, *Clauses 2.1 Quality, 3.6 Quality system and 3.12 Quality manual in Section 2: Terms related to quality* taken from ISO 8402: 1994 and *Clause 5.4.3 Extent of audits* taken from ISO 9004.1: 1994 have been reproduced with the permission of the International Organization for Standardization, ISO. These standards can be obtained from Standards Australia, 1 The Crescent, Homebush, NSW 2140, Australia or directly from the Central Secretariat, ISO, Case postale 56, 1211 Geneva, Switzerland. Copyright remains with ISO.

Finally, I would like to record my appreciation of the encouragement and support of my colleagues at the University of Technology Sydney without which I would not have carried through my writing endeavour.

# PART ONE

# ESTABLISHING A QUALITY SYSTEM

# 1

# Quality and quality assurance

## 1.1 WHAT IS QUALITY?

Quality may mean different things to different people. Some take it to represent customer satisfaction, others interpret it as compliance with contractual requirements, yet others equate it to attainment of prescribed standards. The International Organization for Standardization (ISO) formally defines quality as the 'totality of characteristics of an entity that bear on its ability to satisfy stated or implied needs' (ISO, 1994a). Dr J. M. Juran, an international authority in quality management, perceives quality simply as 'fitness for purpose'. Indeed, a product befitting its intended purpose would satisfy the user's needs and expectations. The crucial point lies in making the purpose clear to all parties involved in the design and production.

In the context of quality management, quality is not an expression of excellence in a comparative sense. It is just an abbreviation for 'desired quality' that should be laid down as explicitly as possible. The supplier (producer), on the one hand, endeavours to attain the desired quality at optimum cost while the customer, on the other hand, requires confidence in the producer's ability to deliver and consistently maintain that quality.

Quality of construction is even more difficult to define. First of all, the *product* is usually not a repetitive unit but a unique piece of work with specific characteristics. Taking building construction as an example, the product can be an entire building, a section of a building or just a prefabricated component that ultimately forms part of a building. Secondly, the needs to be satisfied include not only those of the client but also the expectations of the community into which the completed building will integrate. The construction cost and time of delivery are also important characteristics of quality. All these should be properly addressed in designing the building, and the outcome should be expressed unequivocally in drawings and specifications.

*4   Quality and quality assurance*

The quality of building work is difficult, and often impossible, to quantify since a lot of construction practices cannot be assessed in numerical terms. The framework of reference is commonly the *appearance* of the final product. 'How good is good enough?' is often a matter of personal judgement and consequently a subject of contention. In fact, a building is of good quality if it will function as intended for its design life. As the true quality of the building will not be revealed until many years after completion, the notion of quality can only be interpreted in terms of the design attributes. So far as the builder is concerned, it is fair to judge the quality of his work by the degree of compliance with stipulations in the contract, not only the technical specifications but also the contract sum and the contract period. His client cannot but be satisfied if the construction is executed as specified, within budget and on time. Therefore, a quality product of building construction is one that meets all contractual requirements (including statutory regulations) at optimum cost and time.

## 1.2   QUALITY CONTROL

Quality control refers to the activities that are carried out on the production line to prevent or eliminate causes of unsatisfactory performance. In the manufacturing industry, including production of ready-mixed concrete and fabrication of precast units, the major functions of quality control are control of incoming materials, monitoring of production processes and testing of the finished product.

Before production is commenced, an assessment is made of the minimum quality needed to satisfy the stated requirements and how that quality can be consistently achieved. An example is establishing the target mean strength of concrete on the basis of the specified characteristic strength and the estimated variability. During production, the strength of the concrete is continuously monitored via routine testing and statistical analysis of the test results, so as to detect at the earliest possible moment when either the mean strength or the variability of strength shows a significant change. The control mechanism then goes on to rectify the detected change, thereby preventing a potential problem from developing into a real one.

Very rigid rules for production control may be combined with lenient criteria of acceptance and trivial consequences of noncompliance. Alternatively the producer may be given greater freedom in production, but stringent acceptance criteria are set and severe penalties for noncompliance are imposed. Within the spectrum of possible combinations of production control and acceptance control, there will be an optimum that is the most economical to operate.

In the building industry, it is traditional practice to have separate contracts for design and construction, with the designer also taking up the role of supervision of construction. The quality of the finished works is controlled by way of inspection and testing as construction proceeds. For example, the quality of concrete and other materials on site is judged by random sampling and testing, and a thorough inspection of the finished works is performed without exception before final acceptance. The major drawback of this 'inspectorial system' of quality control is that it identifies the mistakes after the event. Even high strength concrete can be defective if it is not properly compacted and cured, and the potential hazard of steel corrosion will not surface until some years later. Many building defects are covered up during subsequent construction and consequently the quality of the finished works cannot be assessed by final inspection. Unlike consumer goods, defective building work is very difficult, if not impossible, to replace. The client is often left with the patched-up original which will be a source of recurrent trouble and huge expenditure in the years to come.

Regular supervision by the contractor's staff themselves is undoubtedly the key to quality. There are, however, commercial and organizational pressures that often favour speed at the expense of quality. Sometimes poor workmanship is condoned to keep up with expected productivity or just labour. To show commitment to quality, senior management of the company must therefore provide adequate resources on site to avoid anybody cutting corners. Furthermore, a comprehensive record of in-process inspection is essential to ensure that the intended verification is actually done. The extra efforts are managerial in nature and complementary to the operational techniques of quality control in assuring the quality of the product.

## 1.3 QUALITY ASSURANCE

Despite the wealth of site experience accumulated throughout the decades, one in ten building contracts still leads to client dissatisfaction and complaint against the contractor. A survey conducted by the Building Research Establishment in the United Kingdom indicates that 40% of building defects occur during the construction phase (BRE, 1982). In most cases, the defects are found to be the result of:

- misinterpretation of drawings and specifications;
- use of superseded drawings and specifications;
- poor communication with the architect / engineer, subcontractors and material suppliers;
- poor coordination of subcontracted work;
- ambiguous instructions or unqualified operators;
- inadequate supervision and verification on site.

It is obvious that defects arising in construction are mostly caused by poor management and communication. It is preclusive to assume that mistakes appearing on site are actually made on site. These mistakes may be traced back to the purchase of incorrect or incompatible materials and the failure to retrieve the out-dated drawings (Kettlewell, 1990). In other words, site problems can be the consequence of negligence or malpractice in the head office.

Consistent quality can only be achieved when such *avoidable* mistakes are avoided in the first instance. Preventive measures must be taken to minimize the risk of managerial and communication problems. This is the basic concept of quality assurance.

The performance of an individual in an organization could directly or indirectly affect the quality of the finished product. Responsibility for quality therefore stretches from the chief executive right down to the person-on-the-job. If consistent quality is to be assured, all staff in the organization, both in the head office and on site, must:

| | |
|---|---|
| • know what their authorities are: | have appropriate organization structure, clear lines of responsibility and communication; |
| • know what their duties are: | have clear definition and description of duties; |
| • know what to do: | have correct specifications and drawings; |
| • know how to do it: | have proper training, appropriate procedures, ready access to necessary instructions; |
| • want to do it: | have proper motivation; |
| • be able to do it: | have the right resources, plant and materials; |

- know that it is done: have appropriate checking, measurement or testing of products;
- record that it has been done: keep proper records, specified certificates (CIRIA, 1989).

To practise quality assurance, an organization has to establish and maintain a quality management system (usually abbreviated to quality system) in its day-to-day operation. A quality system contains, among other things, a set of documented procedures for the various processes carried out by the organization. Implementing a quality system does not replace the existing quality control functions, nor does it result in more inspection and testing; it just ensures that the appropriate type and amount of verification is performed when and where it is planned to be done. In fact, a quality system embraces quality control as its technical arm. This is why a quality system is sometimes referred to as a QA/QC programme.

In short, quality assurance is oriented towards *prevention* of quality deficiencies. It aims at minimizing the risk of making mistakes in the first place, thereby avoiding the necessity for rework, repair or reject.

## 1.4 IS QUALITY ASSURANCE FOR CONSTRUCTION ?

In the construction industry, quality assurance was first adopted in nuclear installation and offshore works mainly for safety and reliability reasons. Spread of the concepts to conventional types of construction has been gradual but slow. This is because the product of construction is in a sense always unique, unlike consumer goods which are repetitive in nature. The processes of construction involve a variety of professionals and tradesmen with a wide range of skills and level of education. The environment where these processes are carried out is often exposed to aggressive elements. Under such conditions, it is arguable whether the procedures can be standardized at all. Some contractors even think that trying to do so merely implants another layer of bureaucracy in the organization.

Despite the diversity of work handled by a construction company, the *corporate* procedures apply to all projects in varying degrees. Typical examples are tendering, procurement, document control and record keeping. A quality system may be set up to standardize these corporate procedures, with provision for preparation of a quality plan to cover the characteristics and specific requirements of a particular project.

## 8  Quality and quality assurance

It is unfortunate that adoption of quality assurance in the construction industry has been mainly client-led. Realizing that enforcement of the contract in law cannot undo any damage already done[*], a progressive client, when awarding a contract, tends to take into account the contractor's capability to 'do it right first time, every time' – the underlying philosophy of quality assurance. There is a general movement towards making the implementation of a quality system a contractual requirement. Many government bodies responsible for public works and housing have begun to insist on an effective (or even certified) quality system as prerequisite for tendering. Public utilities companies are doing the same thing. Private developers with major projects in planning will follow suit. The basis of competition for business will shift from 'price only' to a combination of price and quality. If a contractor does not want to be excluded from bidding for available work, he should wait no more in establishing a quality system in his organization. Even if such external pressure is not on at the moment, he will be fighting a losing battle against his competitors who have enhanced their productivity through better quality management.

Just to satisfy a condition for tendering or contract may not be the best argument for practising quality assurance, but it is probably the most compelling reason in the first instance. However, the companies that benefit most from quality assurance are those which do so for the purpose of improving their own efficiency (Ashford, 1989). Notable changes are more effective communication, both within the organization and with outside bodies, less disruption of work and reduced spending on rework. These improvements lead to higher productivity on the one hand and client satisfaction on the other.

Of course it costs to implement and maintain a quality system. Significant investment in terms of money and staff time is needed en route to quality assurance, especially for document preparation and staff training. Some people see this as another item of overhead for the company. However, they should not lose sight of the savings that will accrue later with much reduced incidents of rework or reject. The overall quality related costs decrease rapidly as quality awareness among the staff increases.

The actual cost is difficult to gauge and varies substantially with the size of the company and the scope of its operations. A questionnaire survey of contractors in Hong Kong has indicated that the setting up cost ranges from HK$1 million (US$128 000) to HK$3 million (US$385 000) with an

---

[*] In the event of the construction not complying fully with the specifications, the client may be awarded financial compensation but left with the defective works.

average running cost of around 0.2% of contract value (Tam, 1996). An analysis of seven building projects of various sizes in Australia has demonstrated that 'quality does not cost – it pays' (Roberts, 1991). The results of the analysis are summarized in Fig. 1.1 in which the quality related costs are expressed as percentages of the total construction cost. Through the implementation of a proactive quality system that costs about 1% of the project value (the prevention cost), the expenditure as a result of repair etc. (the failure cost) drops from 10% to 2%, representing a saving of 7%. The economic benefit of *preventive* measures is obvious.

Quality assurance has also less obvious benefits. For a contracting company, a well-established quality system is a marketing tool, especially when the company has gone through third party certification. The quality system helps promote the image of the company and provide the much needed competitive edge in a competitive market. Improved market share will more than pay back the investment.

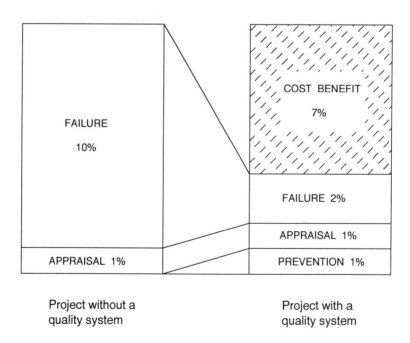

**Fig. 1.1** Implementation of quality management (Roberts, 1991).
*(reproduced by permission of Master Builders Australia)*

*10 Quality and quality assurance*

The quality records generated by the quality system facilitate and strengthen the process of claims and, in case of counter-claims, provide a potential line of defence. Legal costs are minimized, should litigation be required at all.

## 1.5 SUMMARY

- Quality in general terms is 'fitness for purpose', but in building construction it is more appropriately interpreted as compliance with contractual requirements.
- Quality control, such as inspection and testing, does not by itself achieve consistent quality because the intended verification is not always done.
- A system of quality management complements the operational techniques of quality control in assuring quality. The product has 'build-in quality' instead of 'inspect-in quality'.
- Quality assurance by way of a well-structured quality system avoids problems and improves efficiency, resulting in high productivity and client satisfaction. There are also less obvious benefits.

# 2

# Quality system and system requirements

## 2.1 QUALITY SYSTEM

In a manufacturing factory, material and labour are input through a series of processes out of which a product is obtained. The quality of the output is monitored by inspection and testing at various stages of production. Any nonconforming product found is either repaired, reworked or scrapped. In construction, the scenario is similar except that if anything goes wrong, the nonconforming work is very difficult and costly (and sometimes even impossible) to rectify.

To maintain consistent quality of a product, be it a paper clip or a multi-storey building, the organization carrying out the production must have some means to ensure that every time a process is performed the same method is adopted and the same control is exercised. Consistency of operation involves three basic functions: say what you do, do what you say, and record what you have done. This can be achieved by establishing a quality system in the organization and maintaining it to be effective.

A quality system is a framework for quality management. It embraces the 'organizational structure, procedures, processes and resources needed to implement quality management' (ISO, 1994a). The quality system of an organization clarifies the authorities and responsibilities of the staff and their interrelations. It standardizes the administrative and the production procedures. It regulates the conduct of verification activities. In addition, it generates permanent records showing that the prescribed activities of verification have been performed and the required quality has been attained.

The quality system is fully integrated with the operations of the organization as shown in Fig. 2.1. It regulates the conduct of the different processes and prevents side-stepping. If at any time a certain process is found to deviate from the established procedure, the untoward event is

## 12 Quality system and system requirements

reviewed by management and the loophole is plugged to prevent recurrence.

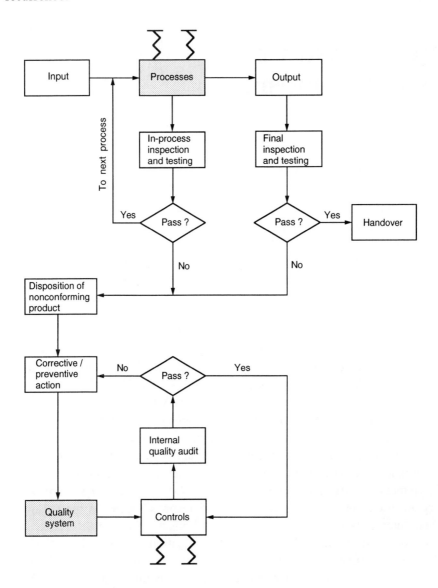

**Fig. 2.1** Integration of quality system with operational processes.

As concerted effort is central to quality assurance, everyone in the organization needs to know not only what he/she is expected to do but also what his/her colleagues are doing. Therefore, the quality system has to be fully documented and readily available in the workplace. Furthermore, a documented quality system renders itself amenable to *controlled* change. In the past, many companies relied on long-serving staff to pass on the traditional practices to newcomers. Such arrangement was workable when things remained static and people were stuck with one employer for long periods. In modern times, technologies are changing at an unprecedented rate and client's demands become more and more sophisticated. Besides, rapid turnover of staff is the norm rather than the exception. Consequently, the quality system will be amended every now and then to keep pace with the evolution, and the system itself must ensure that the changes are conveyed to all those who are affected.

The required documentation of a quality system ties in closely with the basic functions as aforesaid. The top management's commitment to quality is expressed in a quality policy statement. The quality policy is incorporated and expanded in a quality manual which sets out what management requires its staff to do to assure quality. How it is to be done is detailed in a number of quality procedures and work instructions. What is actually done is evidenced in the various written records. The hierarchy of quality system documents is portrayed as a pyramid in Fig. 2.2.

**Fig. 2.2** Pyramid of quality system documents.

Within the organization, the quality system is a management tool, providing assurance that quality control activities have been planned and carried out in full. Outside the organization, the quality system is an objective demonstration of the builder's ability to produce building work in a cost effective way to meet the customer's requirements (CIOB, 1987).

## 2.2 QUALITY SYSTEM STANDARD

A quality system has to cover all the activities leading to the finished product. Depending on the scope of operation of the organization, these activities include planning, design, development, purchasing, production, inspection, storage, delivery and after-sales service. A reference base is required against which the adequacy of a quality system can be judged. Such a reference base is called 'quality system standard'.

In the early days of quality assurance, government defence departments on both sides of the Atlantic had their own standards for assessing the supplier's quality system. Naturally these standards were oriented towards the defence-related industries. With the rapid adoption of quality assurance in the 1970s, other major procuring bodies, such as electricity generating boards and automotive manufacturers, developed different standards to suit their specific requirements. Obviously, the consequence was a proliferation of standards.

There was a need to rationalize the various standards, leading to a unified set applicable to all. In 1979 the British Standards Institution made a major contribution in this direction by issuing the three-part standard BS5750 (BSI, 1979). This standard provided a comprehensive coverage of the key elements that were expected of a quality system. It was most useful for third-party certification (registration). A supplier has to be assessed once to become acceptable to many customers. At the same time, a quality conscious customer could select a certified supplier without carrying out a separate assessment.

The International Organization for Standardization (ISO) went one step further by developing a standard which would be internationally accepted. Indeed, compatibility of quality systems and mutual recognition of certification schemes are essential to further development of business in the global sense. Based on BS 5750 and with input from some twenty countries, the Technical Committee ISO/TC 176 of this international body produced the ISO 9000 family of standards in 1987.

## 2.3  ISO 9000 FAMILY OF STANDARDS

The ISO 9000 family is made up of the following standards:

ISO 9000  Quality management and quality assurance standards (in four parts)
ISO 9001  Quality systems – Model for quality assurance in design, development, production, installation and servicing
ISO 9002  Quality systems – Model for quality assurance in production, installation and servicing
ISO 9003  Quality systems – Model for quality assurance in final inspection and test
ISO 9004  Quality management and quality system elements (in four parts)

ISO 9001, ISO 9002 and ISO 9003 are for contractual, assessment or certification use. These International Standards specify the quality system requirements for three categories of organizations with different scopes of operation.* They represent the international consensus on the essential features of a quality system. They serve as a benchmark for supplier assessment throughout the world. An appropriate one can be invoked in a contract between the customer and the supplier as a contract requirement. On the other hand, ISO 9000 and ISO 9004 are not for contractual purposes; instead, they provide guidance to the use of the other standards in the family.

There are other standards closely related to ISO 9000. These standards each describe a particular aspect of quality management and quality assurance in more details, such as quality planning and quality auditing. Table 2.1 shows a collection of the ISO 9000 family and related standards.

Following its first issue, the ISO 9000 family of standards has been adopted by individual countries as their national standards, sometimes with slightly different titles. For example, in the USA it is known as ANSI/ASQC Q9000 while the Australasian version is AS/NZS 9000. In the United Kingdom, BS 5750 (the predecessor of ISO 9000) has become BS EN ISO 9000. The number of countries accepting the standards has grown very fast. Up to date, over 80 countries have done so. The standards are indeed international.

---

\*   Despite their titles, these standards are not *model quality systems* but *model standards for quality systems*.

*16   Quality system and system requirements*

**Table 2.1(a)**   ISO 9000 family of standards

| | |
|---|---|
| ISO 9000-1:1994 | Quality management and quality assurance standards – Part 1: Guidelines for selection and use |
| ISO 9000-2:1993 | Quality management and quality assurance standards – Part 2: Generic guidelines for the application of ISO 9001, ISO 9002 and ISO 9003 |
| ISO 9000-3:1991 | Quality management and quality assurance standards – Part 3: Guidelines for the application of ISO 9001:1994 to the development, supply and maintenance of software |
| ISO 9000-4:1993 | Quality management and quality assurance standards – Part 4: Guide to dependability programme management |
| ISO 9001:1994 | Quality systems – Model for quality assurance in design, development, production, installation and servicing |
| ISO 9002:1994 | Quality systems – Model for quality assurance in production, installation and servicing |
| ISO 9003:1994 | Quality systems – Model for quality assurance in final inspection and test |
| ISO 9004-1:1994 | Quality management and quality system elements – Part 1: Guidelines |
| ISO 9004-2:1991 | Quality management and quality system elements – Part 2: Guidelines for services |
| ISO 9004-3:1993 | Quality management and quality system elements – Guidelines for processed materials |
| ISO 9004-4:1993 | Quality management and quality system elements – Guidelines for quality improvement |

**Table 2.1(b)** ISO standards related to ISO 9000 family

| | |
|---|---|
| ISO 8402:1994 | Quality management and quality assurance – Vocabulary |
| ISO 10005:1995 | Quality management - Guidelines for quality plans |
| ISO 10011-1:1990 | Guidelines for auditing quality systems – Part 1: Auditing |
| ISO 10011-2:1991 | Guidelines for auditing quality systems – Part 2: Qualification criteria for quality systems auditors |
| ISO 10011-3:1991 | Guidelines for auditing quality systems – Part 3: Management of audit programmes |
| ISO 10012-1:1992 | Quality assurance requirements for measuring equipment – Part 1: Metrological confirmation system for measuring equipment |
| ISO 10013:1995 | Guidelines for developing quality manuals |

The spontaneous response to the ISO initiative is no surprise: no one can afford to be left out in today's international trade. In fact, the ISO 9000 family of standards creates a 'level playing field' where everyone can join. Thousands of companies throughout the world have their quality systems certified to one of these international standards, and the number is fast growing. With such world-wide application, a wealth of experience was soon accumulated. Analysing the feedback from the users, ISO/TC 176 undertook revision of the standards and the second edition was issued in 1994.

The ISO 9000 family of standards is intended for application in all industries. The generic requirements of these standards have to be interpreted in the context of the conventional practices of the specific industry. For this purpose, various industry guides have been published. One such guide is AS/NZS 3905.2: 1997 for the construction industry in Australia and New Zealand (SA/SNZ, 1997).

In the construction industry, ISO 9001 is suitable for companies that are engaged in design-and-construct projects, including domestic house building. However, the majority of building contractors work to a design supplied by the architect / engineer, and their quality system may be modelled on ISO 9002. (In fact, ISO 9002 is a sub-set of ISO 9001 without the requirements on design control.) ISO 9003 is for use where compliance is judged solely by final inspection and test, which is obviously not the case in building construction. Consequently the quality system of a building construction company should conform to either ISO 9001 or ISO 9002, depending on whether design of permanent works forms part of the company's activities.

## 2.4 QUALITY SYSTEM REQUIREMENTS

To cover the full range of activities pertaining to quality assurance, the ISO 9000 family of standards identifies twenty elements of the quality system and sets out definitive requirements appropriate to each. For each element, the required control is prescribed but not the methods by which control can be achieved. It is up to the company's management to establish and maintain documented procedures to ensure that these requirements are consistently met. The requirements in ISO 9001 and ISO 9002 are the same, except that the element of design control is not included in the latter.

The quality system requirements stipulated in ISO 9001 / ISO 9002 are not reproduced here. Instead, they are interpreted, element by element, in the context of building construction. In reading the following sub-sections, you are advised to consult the respective clauses of the standard. You should also refer to the corresponding quality system documents in Chapter 9. A signpost pointing to the relevant document *[in square brackets]* is provided at the end of each sub-section. Cross-reference is also shown in Table 9.1 (page 124).

### 2.4.1 Management responsibility

The responsibility for quality lies with senior management of the company. A quality policy statement, signed by the chief executive, should be issued setting out the company's objectives for quality and the management's commitment to quality. The quality policy is publicized among all staff of the company, e.g. through seminars and indoctrination sessions, so that it is fully understood and followed.

The responsibility, authority and interrelation of staff in the company should be defined and documented. This is usually done by means of an organization chart coupled with position descriptions. For a construction company, the organization chart should preferably show the interrelation between office staff and site staff.

No quality work can result without the provision of adequate resources. This applies to activities in the office as well as those on site. The needs of each operation should be identified beforehand, and appropriate human resources and equipment made available at the right time. This includes training of personnel before they are assigned to specific tasks.

A senior member of staff should be put in charge of quality related matters. This person, bearing the title of quality manager or a similar title, also acts as the management representative in liaison with external parties. He/she has direct access to the chief executive of the company and is authorized to implement and maintain the company's quality system. By necessity, the quality manager is experienced in quality management techniques and knowledgeable in all facades of the company's business.

Senior management should review the quality system at defined intervals, taking into account the outcomes of quality audits (both internal and external) and feedback of clients. The quality management review is to ensure that the quality system continues to be suitable and effective. It also provides an opportunity to modify the quality system to cater for changes in business direction of the company, client needs or government regulations. It is often necessary to conduct the review at two levels: a corporate management review and a project management review. The former is normally carried out annually while the latter is held more frequently, say at half-yearly intervals. *[Quality manual, pp. 129–38; Procedures QP1.1,1.2, pp. 154–7]*

## 2.4.2 Quality system

A company's quality system is its blue-print for quality management. It is a means of ensuring the quality of the product.

The quality system is to be fully documented with documentation usually in three tiers. The first tier document is the quality manual which covers the requirements of the International Standard to which the quality system is to conform. It also incorporates the quality policy, the organization chart as well as an outline of the structure of the documentation of the quality system. The second tier documents consist of a number of quality procedures which are referred to by the quality manual. (The quality procedures may be included in the quality manual, but the single document so formed is in most cases too voluminous to be practical.) The quality procedures are mostly administrative in nature. They are supplemented by work instructions (third tier documents) that define how the activities are performed. The quality procedures and work instructions together ensure consistency of operations.

On the project basis, the quality system is implemented through a quality plan. A quality plan is a document setting out the specific quality activities and resources pertaining to a particular contract or project. Among other things, it makes reference to an assortment of documented procedures of the quality system. It may also contain project-specific procedures or work instructions which only apply to the particular project. While the quality manual is intended to apply across the company's entire organization, the quality plan is prepared specifically for the project in question. It is virtually the quality manual of the project.

Quality planning should be done at an early stage of the project. Before construction work commences, or even before the contract is signed, all special requirements and non-routine processes should be identified and considered. In most cases, the entire project is completely scheduled before ground is ever broken. Time and effort spent on planning will be more than recovered later in the construction process. *[Quality manual, pp. 139–40, 149; Procedure QP2, pp. 158–66]*

### 2.4.3 Contract review

Contract review is a preventive measure to avoid any misunderstanding between the parties to a contract. The first review normally takes place at the tender stage. Before submitting the tender, the contractor should ensure that the requirements of the client are well defined in the drawings, specifications, bills of quantities and other documents, and should clear up any ambiguities and omissions with the architect / engineer. He should also ensure that his organization is adequately equipped, both technically and financially, to undertake the project and meet the specified requirements.

It is not uncommon in a construction project that some detailed drawings or relevant information are unavailable at the time of tender or even when the contract is let. The lack of details does not necessarily hold back the tendering process or signing of the contract, provided that enough information is in hand to interpret the client's requirements with confidence. However, it is wise to agree with the architect / engineer a time frame for the supply of the missing information so that the planned programme of construction can be adhered to.

If the tender is successful, the building contractor should review the contract again before it is formally signed. This time, the objective is to resolve matters arising from changes in requirements since the tender was submitted.

During the progress of the contract, any variation order involving substantial changes in the design or specification of the works, either in kind or in quantity, should be reviewed before acceptance. This is to make sure that the altered requirements can be satisfactorily accommodated with the resources available. It is important to communicate the changes to the people actually doing the work.

The reviews carried out at different stages of the contract are to be recorded. *[Procedure QP3.1, 3.2, 3.3, pp. 167–75]*

### 2.4.4 Design control

Design control is a key element in ISO 9001 but is outside the scope of ISO 9002. The aim of design control is to ensure that the client's requirements, including applicable regulatory requirements, are faithfully translated into drawings and specifications. The design is to be reviewed, verified and validated before being released for use; and this applies also to subsequent changes to the design.

The design of a building is normally prepared by the architect / engineer and supplied to the contractor for implementation. The contractor has no need to include design control in his quality system unless he undertakes design of permanent works as part of his operation, such as in a design-and-construct contract. However, design of temporary works is often the responsibility of the contractor, for which he has to establish appropriate control. This is conveniently dealt with by treating the activity as part of the construction process and consequently subject to process control (section 2.4.9).

### 2.4.5 Document and data control

All documents describing the company's operations are subject to control. Obviously the quality system documents are controlled documents. Other documents belonging to this group are those which control the processes of tender preparation, requisition and procurement, staff recruitment, etc.[*] Document control does not affect the contents of the documents, but

---

[*] These documents are generally compiled in the company's operations manual.

regulates the administrative processes by which a document is created, revised, distributed, retrieved, archived and disposed of. The control mechanism operates through the following:

- review and approval of new and revised document before issue by an authorized person;
- distribution of pertinent issues to prescribed personnel and departments;
- prompt removal (but not necessarily immediate disposal) of obsolete documents to prevent inadvertent use;
- marking of obsolete documents to such effect, if not immediately disposed of.

Contract documents and other project-specific documents should also be controlled. These documents include the conditions of contract, specifications, drawings and bills of quantities, as well as the project quality plan, inspection and test plans, soil test results and other data sheets supplied by the architect / engineer, relevant national standards and codes of practice. If client's approval of the project quality plan is a contractual requirement, the document cannot be released for use until such approval is obtained in writing. Subsequent changes to the plan are to be reviewed and approved as the original document.

A simple way to exercise document control is to establish and maintain a master list of all controlled documents, identifying for each document the current revision status and date of issue. An example of a master list is shown in Fig. 2.3.

Anyone making use of a controlled document should check the currency of the copy in hand by referring to the master list or a controlled copy of the document. Controlled copies are distributed to key members of staff who will place them at convenient locations (including site offices) for other staff to consult. A distribution list of controlled copies is maintained so that, when a revision is made, the recipients are supplied with the amended version and asked to return the out-dated one. To differentiate a controlled copy from other copies, it is usually printed on special paper that is subject to restrictive use. The distinctive feature of the paper may be a large company logo or such mark as 'controlled copy' pre-printed in colour or as a watermark across the page. An example of a controlled copy is shown in Fig. 2.4.

| Doc. No. | Doc. Type | Document title | Issue No. | Revision No. | Date |
|---|---|---|---|---|---|
| QM01 | Manual | Quality manual | 1 | 1 | 5/3/97 |
| OM01 | Manual | Operations manual | 2 | 3 | 12/2/96 |
| SM01 | Manual | Safety manual | 1 | - | 12/5/95 |
| ... | ... | ... | ... | ... | ... |
| QP01-1 | Procedure | Corporate management review | 1 | - | 12/12/96 |
| QP01-2 | Procedure | Project management review | 2 | - | 18/2/97 |
| QP02-1 | Procedure | Preparation of project quality plan | 1 | - | 12/12/96 |
| QP03-1 | Procedure | Contract review | 2 | - | 18/2/97 |
| ... | ... | ... | ... | ... | ... |
| QF03-1 | Form | Checklist for contract review | 2 | - | 18/2/97 |
| QF10-1 | Form | Material receipt and issue | 1 | - | 12/12/96 |
| QF14-1 | Form | Corrective action request | 1 | - | 12/12/96 |
| ... | ... | ... | ... | ... | ... |
| IN03 | Work instruction | Preparation and approval of purchase order | 2 | - | 9/11/94 |
| IN08 | Work instruction | Inspection of welding | 1 | - | 12/11/93 |
| IN15 | Work instruction | Handling of hazardous materials | 2 | - | 3/12/96 |
| ... | ... | ... | ... | ... | ... |

**Fig. 2.3** Example of master list of controlled documents.

**Quality Construction Ltd.**
**Building Contractors**

**QP 5.2 : PROCEDURE FOR CONTROL OF DOCUMENTS FOR SPECIFIC PROJECT**

| | |
|---|---|
| 1 | **Purpose** |
| | To control the processes of approval, revision, issuance, distribution and removal of documents pertaining to a specific project |
| 2 | **Scope** |
| | Applicable to the conditions of contract, specifications, contract drawings, bills of quantities and other documents originated from the client, standards and codes of practice released for site use, project quality plan, project-specific procedures, work instructions, working drawings, site investigation reports and miscellaneous data sheets |
| 3 | **Person responsible** |
| | Contracts Manager / Project Manager / Quality Assurance Officer |
| 4 | **Procedure** |
| 4.1 | The Contracts Manager is responsible for control of documents and drawings supplied by the client. For each project, he establishes and maintains a distribution list of contract documents and drawings using Form QF5-2. |
| 4.2 | When a revised document / drawing is received from the client, the Contracts Manager records it on the list before distribution and ensures prompt return of the document / drawing which is made obsolete. The obsolete document / drawing is stamped as 'SUPERSEDED' before storage. |
| 4.3 | With documents and drawings prepared internally for the project, such as the project quality plan, technical procedures, work instructions for special tasks and working drawings, the Project Manager (or the Engineer as appropriate) reviews and approves such document / drawing before release for use. An approved document / drawing carries the signature of the authorized person. Changes to a document or drawing are reviewed and approved by the same person or the current incumbent of the same position. |
| 4.4 | On site, the Quality Assurance Officer is responsible for document control. He establishes and maintains a register of documents and drawings kept in the site office using Form QF5-3. Controlled copies of such are issued to those persons (including subcontractors) who need them for their operation. Distribution and retrieval of controlled copies are recorded on Form QF5-2. The disposition of the superseded copies is at the discretion of the Project Manager. |

*Quality Procedure QP5.2*  *Page 1 of 2*
*Issue 1, 12/12/96*

**Fig. 2.4** Controlled copy of controlled document.

It is important to pass on the up-to-date issue of contract documents, in particular drawings, to those people who need them to perform their work. It is equally important to remove the invalid and/or obsolete documents from the workplace. If a superseded document is retained for future use, such as for adjustment of the contract sum, it should be properly identified and separated from the current issue. *[Procedures 5.1, 5.2, pp. 176–82]*

## 2.4.6 Purchasing

In the construction industry, purchasing refers not only to procurement of supplies but also to provision of labour and other services. Nowadays there is a tendency for the building contractor to take up a management role, employing a team of subcontractors to carry out the construction activities. (The subcontractors in turn may rely on other subcontractors to work for them.) As the main contractor is contractually responsible to the client for the subcontracted work, it is to his advantage and protection to verify that any supplier of material or labour is capable of providing the right product or service at the right time. Commissioning a consultant to design temporary works is also a form of subcontracting. The bought-in design is therefore treated as a product provided by the consultant.

In respect of purchasing, the main contractor acts as the purchaser (customer) and needs to be assured by the supplier / subcontractor that the intended quality be achieved in the supply or service delivered, just as his organization be required to provide confidence to the client about the quality of the final product – the finished works. Control of purchasing and subcontracting varies with the type of supply or service and its impact on the quality of the finished product. The degree of control should increase with the risks and consequences of nonconformance as indicated by the complexity of the activities involved, the stringency of the acceptance criteria and the cost of rectification of nonconforming work. For the provision of a major item or service, the main contractor should look for evidence of the subcontractor's capability to carry out the subcontract satisfactorily. For a minor item, less stringent control is often adequate.

It is obvious that orders should only be placed with suppliers/ subcontractors who are competent and resourceful, and that purchasing documents (including subcontracts) should be clear and definitive. For satisfactory purchasing, there are two prerequisites, i.e.

- selecting a supplier / subcontractor capable of providing the supply or service of the intended quality;
- specifying the required supply or service in sufficient details to avoid ambiguity.

The capability of a subcontractor for quality may be judged on the basis of available resources (both labour and plant), past performance, and the quality system if there is one implemented. Should the subcontractor satisfy the pre-set criteria in these measures, the main contractor may rest assured that satisfactory performance is most likely to ensue. Otherwise, the main contractor has to provide extra resources for supervision and verification or even to embrace the activities of the subcontractor in his own quality system. With multi-layered subcontracting, which is not uncommon in modern construction, the control efforts must extend far enough to be effective. This supportive role is an indirect cost to the service purchased and should be taken into account in deciding on the successful bid.

Therefore, before employing suppliers / subcontractors, the main contractor is obliged to

- evaluate the ability of the potential suppliers / subcontractors to meet the stated requirements including quality assurance requirements;
- register those pre-qualified on a list of acceptable suppliers / subcontractors for the particular supply or service;
- award purchase orders or subcontracts only to acceptable suppliers / subcontractors;
- re-evaluate the suppliers / subcontractors on the list at regular intervals, removing whichever does not measure up to expectation.

A subcontract should clearly specify the expected quality of the service and the quality assurance measures required. In the same way, a purchase order is to contain detailed description of each item of material or equipment to be supplied including inspection and testing arrangements where appropriate. Furthermore, all purchasing documents should be reviewed and approved prior to release.

In a building contract, the client often reserves the right of verification. A typical example is the right to take samples of concrete, steel and other construction materials for compliance testing. Provisions should be made in the purchase orders and subcontracts for the architect / engineer to check the quality of the supply either at the supplier's premises or on site.

However, such verification does not relieve the main contractor of any of his contractual obligations. *[Procedures 6.1, 6.2, pp. 183–90]*

### 2.4.7 Control of customer-supplied product

It is not uncommon for the client of a building construction project to provide certain materials or products for incorporation into the finished works. Examples of customer (or client) supplied product are materials that the client trades in, products from his own factory, mechanical plants to be installed in the permanent works, and logos or other signs of distinction. On receipt, these items should be verified as to kind and quantity, properly identified and stored to prevent misuse or deterioration. Any item that is demonstrably nonconforming should be referred to the client or his representative for instruction of disposition. *[Procedure 7, pp. 191–2]*

### 2.4.8 Product identification and traceability

Product identification in building construction refers mainly to prefabricated components. These include precast concrete units and semi-assembled steelwork. After the units have passed the appropriate inspection and/or testing, they should be made identifiable as to batch or series. Such markings can simply be painted on the units or indicated with stick-on labels. For mass production, even the use of a bar-coding system may be considered (Baldwin, Thorpe and Alkabi, 1994).

Where traceability is required by the contract, it must be made possible to relate each part of the construction / installation work back to the particular batch of material or component used. On a building site, truck-loads of concrete are traceable via the delivery dockets. Traceability of other materials may be established through an inventory system. Mechanical and electrical appliances installed can easily be identified by their serial numbers, and small items such as pumps and light fittings can be labelled. To the extent required, records are maintained of each delivery and the location of its use in the building. *[Procedure 8, p. 193]*

### 2.4.9 Process control

Construction processes, including erection of temporary works, precasting and prefabrication as well as installation of appliances, should be identified, planned and scheduled well ahead of time. This is normally done soon after the project is commissioned. A construction programme for the entire project, together with the equipment and manpower requirements, is developed long before construction commences. (This may often be facilitated by applying the computerized critical path method.) The outcomes of construction planning are incorporated in the project quality plan.

Apart from developing the overall schedule of work early in the project, it is required to plan the site operations in detail at regular intervals and ensure that adequate resources, such as equipment and manpower, are provided at the appropriate time. It is also necessary to monitor the progress of work on site. A bi-weekly bar-chart, such as the chart on page 196, would be convenient and effective. It shows the schedule of activities over the period and the actual progress achieved.

The construction processes should be carried out by labour with appropriate skill, using suitable and well-maintained equipment, and under regular supervision. With multi-layers of subcontractors working on site, difficulties are often encountered in satisfying this requirement. The problem is aggravated by acute shortage and high mobility of labour during a period of building boom. However, this is not an excuse for bypassing the quality system.

Provision of documented procedures or work instructions is essential for non-routine and unconventional processes. These documents have to be prepared specifically for the project. If the work is subcontracted, the work instruction is issued to the subcontractor unless he has his own work instruction which is considered to be suitable for the purpose.

The importance of regular supervision cannot be over-emphasized. Satisfactory outcome of many a process in construction can only be assured by monitoring the process throughout its course. Typical examples are soil compaction and concrete work. Some processes such as prestressing and installation of ground anchors are not verifiable by subsequent inspection and testing; for such special processes only qualified operators should be assigned to the job. This applies to direct labour as well as subcontracted labour.

Design of temporary works, if performed by internal staff, is a special process which should be assigned to someone qualified to do so. The design work is normally done by an engineer either in the office or on site, and the design output is reviewed by his superior before release.

Process control also includes 'suitable maintenance of equipment to ensure continuing process capability'. This virtually calls for preventive maintenance of all equipment used for construction. Most maintenance activities on site are simple and routine, and may only involve regular lubrication by the operator. However, for sophisticated equipment, a maintenance programme has to be planned, carried out on schedule and recorded.

Process control is closely linked with inspection / testing. While process control prevents sidetracking of the established procedures, inspection / testing verifies that the required quality is actually obtained. To make process control effective, every witness point in an inspection and test plan should be observed, and work should not proceed beyond a hold point without approval by the authorized person. This is especially important for control of subcontracted work where the responsibility for regular supervision now lies with the subcontractor with the contractor playing a monitoring role. *[Procedure 9, pp. 194–6]*

### 2.4.10 Inspection and testing

Inspection and testing required for a project should be indicated in the project quality plan. On a construction site, inspection and testing is carried out at three stages:

- on receipt of purchased or subcontracted items or service;
- during a construction process in which an in-process check is conducted before proceeding to the next step;
- before final delivery or handover of the finished works.

Materials, components and appliances received on site are subject to receiving inspection and/or testing. The amount and nature of checking required vary with the degree of control exercised at the supplier's premises and the recorded evidence of conformance. Where an incoming item is released for urgent use prior to verification, the location where it is used or installed should be recorded.

During construction, inspection and testing should be carried out progressively to ensure that any defective work is not built upon or covered over. The requirements for in-process inspection and testing are usually documented in the inspection and test plans (ITPs) which form part of the quality plan. An ITP lists in sequence the activities involved in a process, specifies the checks or tests to be performed and the acceptance criteria, indicates the hold points when verification of quality is a prerequisite to

## 30 Quality system and system requirements

continuation of work, and identifies the authority of approval at each hold point. There are many construction and installation processes for which ITPs have to be prepared. Some examples are excavation and earthwork, piling, concreting (including precasting), structural steelwork, brickwork and blockwork, roofing and cladding, plumbing and drainage, installation of mechanical and electrical services.

Inspection and testing of construction work is traditionally the responsibility of the architect / engineer acting as the client's representative. In the spirit of quality assurance, inspection and testing is carried out mainly by the contractor. Nevertheless, the architect / engineer can exercise control through the hold points and witness points of the ITPs. He may also retain the authority of approval at the hold points where appropriate.

Like the architect / engineer, the contractor monitors the quality of subcontracted work through strategically located hold points. His inspection and test plan should appropriately interface with the subcontractor's if there is one.

Before handover of the finished works or part of it, the contractor always arranges for a final inspection in the presence of the architect / engineer. The final check should include verification that all receiving and in-process verifications required by the quality plan have indeed been fully, correctly and satisfactorily accomplished. This is very important as it is often impossible to judge the quality of the finished product by final inspection alone.

Throughout a construction process, records of inspection and testing are to be maintained. This is conveniently achieved by signing off the inspection and test plan at the various hold / witness points by the designated inspector. Any other method of record keeping may be used provided that it shows clearly whether the respective stages of inspection and testing have been passed. If a nonconformity is discovered during inspection and testing, the part of works in question will be subject to 'control of nonconforming product' (section 2.4.13), and the records should refer to the document regarding the disposition of the nonconforming product. *[Procedures 10.1, 10.2, 10.3, pp. 197–205]*

### 2.4.11   Control of inspection, measuring and test equipment

In building construction, equipment used for inspection, measuring and testing includes surveying instruments, prestressing jacks, torque wrenches, weigh-bridges, various testing devices, weighing scales and measuring tapes. Such equipment, both self-owned and hired, should be properly maintained and calibrated at regular intervals as recommended by the respective manufacturers. A schedule of calibration is established and the records of calibration maintained. (With measuring tapes, it is normally sufficient to check for wear and tear every now and then.) The equipment should be handled, preserved and stored in such a way as to maintain accuracy and fitness for use.

A piece of calibrated equipment should bear a sticker or tag showing the ranges calibrated and the date of calibration. In addition, it is advisable to put on the label the date when the next calibration is due. The equipment should be safeguarded by suitable means from adjustments which would invalidate the calibration setting.

Equipment calibration is often carried out by various specialist firms. If this is performed by internal staff, it is necessary to ensure that the staff are knowledgeable and experienced in the job and work instructions are provided for the purpose. Furthermore, all measuring masters used must be calibrated and traceable to national / international standards.

When a piece of equipment is found to be defective or out of calibration, it should be clearly marked to avoid inadvertent use. At the same time, the validity of measurements made with the equipment since its last calibration is assessed and documented.

The control measures should extend to cover equipment belonging to subcontractors. A subcontractor with equipment on site is required to provide evidence of regular calibration of the equipment. Alternatively, the contractor may include the subcontractor's equipment in his schedule of calibration and instructs the subcontractor to arrange for calibration whenever it is due. *[Procedure 11, pp. 206–8]*

### 2.4.12 Inspection and test status

On the production line, workers must be able to distinguish a product (or semi-product) which is satisfactory from another which is not. Likewise, in building construction, anything that will finally become part of the completed building should have its inspection and test status identified. This includes bought-in materials and components, as well as the contractor's own product either on or off site. The identification is maintained throughout the construction process, from receipt of incoming supplies to handover of the finished works or part thereof.

While manufactured goods can be marked, stamped or tagged to indicate conformance or nonconformance, building works can hardly be labelled in the literal sense except for precast concrete units and steel sub-assemblies. Instead, the inspection and test status is shown in the various inspection and test records. With materials and components received on site, records of receiving inspection and testing provide documented evidence of quality. With building works, records of in-process inspection and testing virtually become a register of the stage-by-stage acceptance of the works. *[Procedure 12, p. 209]*

### 2.4.13 Control of nonconforming product

Nonconforming product or, more appropriately nonconforming work, may arise at different stages of construction. The nonconformities range from minor discrepancies such as sand-streaking of the concrete surface or using a paint of the wrong colour, to major mistakes such as incorrect level of a floor or non-verticality of a wall.

With in-process inspection and testing properly and conscientiously performed, it should be able to discover any nonconformity as soon as it exists. The next step is to identify the nonconforming work so as to avoid it being covered over or built upon. The method to do so depends on the kind of work. In any case, the identification should remain in place until a decision is made regarding the disposition of the nonconforming work. Should it be inconvenient or impracticable to mark the nonconforming work as such, it is necessary to alert the respective functions concerned (including subcontractors) of its existence and location.

The nonconforming work should be reviewed as soon as possible after it is detected. In the review process, the extent of nonconformance and its bearing on the quality of the finished works as a whole are evaluated. The review is conducted by a person who has executive power to take corrective action. This person is normally the project manager or the site agent, but for minor nonconformities the general foreman may act with delegated authority.

Based on the outcome of the review, the disposition of the nonconforming work is worked out, which ranges from qualified acceptance to reconstruction. Sometimes alternative criteria of acceptance are put forth to the architect / engineer for consideration. Any concession granted may come with an adjustment of the contract sum as a penalty. Usually some further investigation and/or remedial work has to be done before the matter is resolved. If the nonconforming work is repaired, it should be re-inspected and/or tested. Records must be kept of any repair or concession granted.

Should the nonconformity exist in subcontracted work, a nonconformance notice is generally issued to the subcontractor indicating the clause(s) of the subcontract violated and/or the specifications not satisfied, the suggested remedial measures and the deadline for implementation. In this respect, it is important that the contractor provides the subcontractor with every assistance. A joint effort often results in speedy solution of the problem and harmonious relationship between the two parties.

Client complaints may be considered as nonconformities and handled in a similar manner. *[Procedures 13.1, 13.2, 13.3, pp. 210–21]*

### 2.4.14 Corrective and preventive action

In the course of reviewing nonconforming work, or handling a client complaint, the cause of the incident and the situation leading to it are usually revealed. A construction activity might have deviated from the contract drawings / specifications or the workmanship might have fallen short of the specified or implied level of standard. Corrective action is necessary to eliminate the cause so as to avoid recurrence of the untoward event. This may involve amending a documented procedure or work instruction, providing additional resources or training the operational staff. In contemplating the appropriate action, account should be taken of the magnitude of the problem and the risks involved.

Prompted by the actual nonconformity discovered, the investigation is usually extended to similar situations in which potential nonconformities exist. (Potential nonconformities may also be revealed during quality audits.) Appropriate steps have to be taken to prevent a potential problem developing into a real one.

Corrective and preventive action should be implemented by the authorized person(s) following an established procedure. Such action may be immediate or long-term in nature. An immediate action is applied to resolve the problem in hand. Its implementation is conveniently verified by signing off the nonconformance record. However, nonconforming work is often the result of contravention of certain documented procedure or inadequacy in the procedure itself. This can hardly be rectified without a long-term action involving staff training, equipment upgrading and changes to the working environment. Its effectiveness is monitored through the feedback of the functions concerned in due course. *[Procedure 14.1, 14.2, pp. 222–8]*

## 2.4.15 Handling, storage, packaging, preservation and delivery

In the context of building construction, this quality system requirement refers mainly to handling and storage of building materials and protection of prefabricated units in transit. From receipt of building materials through to handover of finished works, the contractor should exercise control to prevent damage and deterioration. Storage areas on site and special handling facilities to be provided are normally indicated in the project quality plan. Methods for authorizing receipt to and dispatch from the storage areas are stipulated either in a documented procedure or in the quality plan. For some materials such as cement which are likely to deteriorate with time, the intervals of quality checks are also indicated. *[Procedure 15, pp. 229–31]*

## 2.4.16 Control of quality records

Quality records are intended to demonstrate conformance to specified requirements and effective operation of the quality system. Accordingly, these records fall into two categories which are filed separately.

- Project-specific records:
  These records, which include pertinent records from the subcontractors, provide evidence showing that the required standards of materials and workmanship have been attained.
- System-related records:
  These records should indicate that incidents of non-conformance and client complaints diminish in number with maturity of the system.

The retention period of project-related records is often a matter of contention. To avoid controversy, this period is preferably specified in the contract or indicated in the quality plan. Generally speaking, it is to the advantage of an organization to retain all relevant records for the period of liability, which is normally not less than seven years. Such records may be used, if necessary, as evidence in litigation.

Quality records may be in the form of hard copy or electronic media. Proper maintenance of quality records is an important aspect of a quality system. These records are a prime target of quality auditors, both internal and external. They should be stored in such a manner as to facilitate retrieval yet preventing unauthorized access. Hence, they should be suitably identified, indexed, filed and placed under the control of designated personnel. *[Procedure 16, pp. 232–3]*

## 2.4.17 Internal quality audits

To ensure that what is intended is actually done in the way as prescribed, all activities affecting quality are to be verified regularly. This is achieved via internal quality audits. These audits constitute a self-checking process of the quality system within the organization. The outcome provides evidence that the quality system is, or is not, functioning effectively. Deficiencies in the quality system are often unveiled during the audit, thus paving the way for improvement.

Internal quality audits must be planned and implemented in accordance with established procedures. A schedule of audits, normally on a yearly basis, is prepared, covering all activities related to quality. Some activities are to be audited more frequently than others because they have a more important bearing on the quality of the product or they involve a higher risk of nonconformance.

To obtain an unbiased feedback of the quality system, internal quality audits should be carried out by personnel who are not connected in any way with the activities being audited. An audit normally includes examining records, interviewing operators and observing the work being performed. How revealing an audit would be is dependent on the skills of the auditor(s) in these techniques. Therefore it is necessary to give the auditors appropriate training before assigning them to the task.

Results of the audits should be recorded and brought up in the periodic management review. In the meantime, they should be conveyed to the management staff of the activities concerned so that timely corrective action can be taken on quality deficiencies found. Except for minor deficiencies, a follow-up audit is conducted subsequently to verify that the corrective action has been implemented and that it is effective. *[Procedure 17, pp. 234–40]*

## 2.4.18 Training

All personnel carrying out quality-related functions should have the skills and experience necessary for the tasks. It is the responsibility of management to identify training needs and to provide appropriate training. This is particularly relevant to new staff who may not possess fully the education, training and/or experience prescribed for the job at the time of recruitment. Training needs also arise when existing staff are assigned to tasks different in nature from what they are used to.

Training is also essential in the use of quality procedures and work instructions, including the correct way of filling in the related forms. Such training is preferably made a regular event. Not only do new recruits require indoctrination in the company's quality policy and procedures, but also existing staff need to be given a booster dose of quality awareness every now and then.

Training programmes range from on-the-job training to short courses offered by professional bodies. There should be some means of assessing the skills of the trainees completing the training programme. For certain trades, such as welding and nondestructive testing, the level of competence is normally demonstrated by certification.

Records of training should be maintained. This is conveniently done by updating the personal record to include the achievement attained in the training programme. *[Procedures 18.1, 18.2, 18.3, pp. 241–6]*

## 2.4.19 Servicing

Servicing here means regular maintenance of a product so that it continues to function effectively. This is usually not applicable to building construction unless the contractor undertakes, under the same or a separate contract, ongoing maintenance of the building after handover. Minor repair work performed during the defects liability period of the contract should not be confused with regular maintenance. It is, in fact, part of the construction process and therefore subject to process control.

## 2.4.20 Statistical techniques

Statistical techniques are useful tools for data collection and analysis in the control of a manufacturing process, such as production of concrete, bricks, precast units and pipes. In a building construction project, statistical techniques are used mainly in analysing material test results and plotting control charts. If such techniques are used, their application should be implemented and controlled by documented procedures.

## 2.5 SUMMARY

- A quality system is a framework for quality management. Its functions are to clarify responsibility and authority of staff and their interrelation, to rationalize the administrative and production processes, and to generate permanent records of verification of quality. The quality system is fully documented in a quality manual and a number of quality procedures.
- A quality system standard constitutes a reference base against which the adequacy of a quality system can be judged. The most widely adopted quality system standards are the ISO 9000 family. These International Standards are most useful for third-party certification (registration), especially when international recognition is intended.
- ISO 9001, ISO 9002 and ISO 9003 specify the quality system requirements for three categories of organizations with different scopes of operation. ISO 9001 provides full coverage of the key elements that are expected of a quality system. ISO 9002 is actually a sub-set of ISO 9001 without the requirements for design control. This would be comprehensive enough for a traditional building construction company which works to the design supplied by the architect / engineer. ISO 9003 is for use when conformance to specified requirements is solely based on final inspection and test, which is obviously not the case in building construction. There are other standards in the same family which provide guidance in the interpretation and implementation of the above-mentioned standards.
- The quality system requirements in the context of building construction are summarized in Table 2.2.

## 38  Quality system and system requirements

**Table 2.2**  Summary of quality system requirements

| ISO 9001 Clause No. | Quality system element | Quality functions required |
|---|---|---|
| 4.1 | Management responsibility | • Define, document and publicize quality policy.<br>• Define and document responsibility, authority and interrelation of staff.<br>• Identify and provide adequate resources.<br>• Appoint quality manager.<br>• Review quality system at regular intervals. |
| 4.2 | Quality system | • Establish, document and maintain quality system.<br>• Prepare and effectively implement documented procedures.<br>• Define and document how quality planning is conducted for a project or contract including preparation of a quality plan. |
| 4.3 | Contract review | • Review tender before submission.<br>• Review contract before signing.<br>• Review variation order before acceptance and transfer amended requirements to functions concerned. |
| 4.4 | Design control | • Plan design activities.<br>• Identify and review design input.<br>• Review, verify and validate design output. |
| 4.5 | Document and data control | • Review and approve documents prior to issue.<br>• Review and approve document changes prior to issue.<br>• Control distribution and updating of documents. |

**Table 2.2 (contd)** Summary of quality system requirements

| ISO 9001 Clause No. | Quality system element | Quality functions required |
|---|---|---|
| 4.6 | Purchasing | • Evaluate and select subcontractors on basis of capabilities for quality.<br>• Exercise appropriate control over subcontractors.<br>• Review and approve purchasing documents (including subcontracts) prior to release.<br>• Specify arrangements for verification and product release of subcontracted product or work at subcontractor's premises if required.<br>• Allow the client or his representative to verify subcontracted product or work at the contractor's / subcontractor's premises where specified in contract. |
| 4.7 | Control of customer-supplied product | • Control verification, storage and maintenance of customer-supplied product. |
| 4.8 | Product identification and traceability | • Identify material and semi-finished product (e.g. prefabricated units) from receipt and during all stages of production, delivery and installation where appropriate.<br>• Provide unique identification of individual product or batches where specifically required. |
| 4.9 | Process control | • Identify, plan and control production, installation and servicing processes, including provision of documented procedures and suitable equipment.<br>• Assign qualified operators to carry out special processes. |

**Table 2.2 (contd)** Summary of quality system requirements

| ISO 9001 Clause No. | Quality system element | Quality functions required |
|---|---|---|
| 4.10 | Inspection and testing | • Conduct receiving inspection and testing of incoming materials and components.<br>• Conduct in-process inspection and testing of semi-finished work in accordance with quality plan.<br>• Conduct final inspection and testing of finished work in accordance with quality plan.<br>• Maintain signed-off records of inspections and tests. |
| 4.11 | Control of inspection, measuring and test equipment | • Use inspection, measuring and testing equipment capable of necessary accuracy and precision.<br>• Calibrate the equipment at prescribed intervals, or prior to use, and indicate its calibration status.<br>• Review previous results when the equipment is found to be out of calibration. |
| 4.12 | Inspection and test status | • Indicate by suitable means the conformance or nonconformance of product or work with regard to inspection and tests performed. |
| 4.13 | Control of nonconforming product | • Identify, and segregate when practical, any nonconforming product or work.<br>• Review and dispose of the nonconforming product or work by an authorized person.<br>• Inspect and/or test the product or work again after repair. |

**Table 2.2 (contd)** Summary of quality system requirements

| ISO 9001 Clause No. | Quality system element | Quality functions required |
|---|---|---|
| 4.14 | Corrective and preventive action | • Investigate cause of nonconformities, including client complaints.<br>• Take corrective / preventive action to eliminate cause / potential cause of nonconformities.<br>• Implement and record changes to documented procedures resulting from corrective / preventive action.<br>• Ensure that corrective / preventive action is taken and that it is effective. |
| 4.15 | Handling, storage, packaging, preservation and delivery | • Establish methods of handling product that prevent damage or deterioration.<br>• Use designated storage areas to prevent damage or deterioration.<br>• Assess condition of product in stock at appropriate intervals.<br>• Protect product during delivery. |
| 4.16 | Control of quality records | • Retain quality records for prescribed period.<br>• Maintain quality records in such way that they are identifiable, retrievable and secured against damage, deterioration or loss. |
| 4.17 | Internal quality audits | • Plan and schedule internal quality audits.<br>• Assign independent personnel to carry out internal quality audits.<br>• Conduct follow-up audits if necessary. |
| 4.18 | Training | • Identify training needs of staff.<br>• Provide training required. |
| 4.19 | Servicing | • Verify that servicing meets specified requirements. |
| 4.20 | Statistical techniques | • Identify the need for statistical techniques in quality control.<br>• Implement and control the application of statistical techniques. |

# 3

# Project quality management

## 3.1 PROJECT QUALITY PLANNING

The construction site is the 'production line' of the contractor. It is here that a substantial part of the quality system is implemented. In fact, Clauses 4.7 to 4.15 of ISO 9001 / ISO 9002 are applicable mainly to site operations.

In the construction industry which is project-oriented, a contractor may be working at any one time for a number of clients on a variety of projects. Each project has its own characteristics and requirements. The contractor's quality system must be capable of being tailored to meet the specific needs of the project. The mechanism for doing so is the project quality plan, simply called the quality plan. Preparation of a quality plan ensures that specific requirements of the project are considered and planned for. It also demonstrates to the client how the requirements will be met. Details of a quality plan are given in section 3.2.

Quality planning is a crucial step. It should be done well before construction work is due to commence. It is amalgamated with the traditional project planning in such activities as nomination of subcontractors and suppliers, determination of construction methods, construction programming, site layout, identification of manpower requirements and training needs, material and plant acquisition, etc. For example, subcontractors are to be selected from the current lists of acceptable subcontractors and, where specified in the contract, only those subcontractors operating an effective quality system are employed.

The foremost task of quality planning is to deliberate on the construction methods and any restrictions thereon. In one project it is expedient to use slipforming while in another project off-site precasting is an economical alternative. For these special processes, technical procedures or work instructions may not exist in the quality system documentation and must be developed specifically for the project. Even with conventional methods of construction, some established procedures may have to be modified to cater for certain site restrictions such as noise abatement, environmental protection and traffic control.

While deciding on the construction methods, work out the sequence of operations and draft the construction programme (and sub-programmes). For a sizable project, it is often worthwhile to make use of critical path techniques for which computer programmes are commercially available. In the planning process, consider carefully the resource requirements, both in plant and manpower, versus the projected resource availability at the crucial periods of the project. Think of any factor that would affect the progress of construction. For example, if the project requires special skills that are not available at the moment, arrange for such skills to be acquired through training of existing staff or recruitment of new staff at the appropriate time. The construction programme will, of course, be updated as construction proceeds, but comprehensive scheduling at an early stage paves the way for on-time delivery of the final product.

There may be more than one contractor working on the same project at the same time. For a sizable housing scheme, different contractors may be building different blocks of apartments simultaneously. Or, one contractor may be employed for the building work, another for the access road work, yet another for the sewage pipeline and treatment work. Careful planning of site access and layout is essential to ensure mobility and efficiency during construction.

As a substantial proportion of work is subcontracted, coordination of subcontractors and suppliers is central to smooth running of the project. Subcontractors of different trades are required to start work at different stages of construction. Similarly, suppliers of different materials are required to supply at different times. A schedule (or schedules) should be drafted based on the construction programme, showing respectively the dates that subcontractors are expected to join in and the suppliers to make delivery. This is to ensure that subcontracts and purchase orders are placed in time for the construction work to progress without delay.

Another major task of quality planning is to identify the inspection and testing needs of both materials received and works constructed. Materials used for the project, including components and appliances to be incorporated into the building works, may be categorized according to the level of verification required. For a building project, a lot of proprietary materials are used, e.g. tiles, carpets, sanitary fitments, paints, etc. Samples of these materials are to be submitted to the architect / engineer for approval, but no verification other than receiving inspection is necessary. There are materials such as movement joints, bearings and cables which require sophisticated testing to evaluate their performance. These materials are normally tested in the place of production and test certificates are supplied with the products. Cement, concrete, reinforcing bars, bricks, pipes, etc. form another category of material. These materials are normally sampled on site for standard tests. The type and frequency of test are extracted from the specifications or the relevant national standard for the material. If possible, an accredited laboratory is commissioned to undertake the test in due course.

Inspection and testing needs of building works are covered by inspection and test plans, usually abbreviated as ITPs. It should be decided at the quality planning stage which processes require an ITP. Generally an ITP is required if a construction or installation process involves in-process verification. This applies to work by direct labour as well as subcontracted work. During preparation of an ITP, e.g. for pile testing, identify any special measuring or testing equipment required and arrange for its acquisition in time. Details of ITP are given in section 3.3.

In a building construction project, it is not uncommon to have a number of subcontractors working under the auspices of the contractor. In this respect, the contractor acts as the client and the subcontractors as the suppliers of services and/or materials. Most subcontractors are small companies undertaking small packages of work for the project. The contractor's quality management often has to embrace the activities of the subcontractors. It is advisable to demand an inspection and test plan from each subcontractor for his part of work. This is to encourage the subcontractor to exercise quality planning and control.

In the spirit of quality assurance, the responsibility for process control rests essentially with the contractor, with the architect / engineer taking a monitoring role. Adequate supervisory staff should be planned for and provided at appropriate times. The site supervisors, e.g. the site engineer and the foremen, inspect the work in progress, then either permit the work to proceed, or identify any nonconforming work and report it to the project manager for disposition. Written records are kept of all verification activities. This is conveniently done by signing off the respective inspection and test plans.

The quality plan is the linchpin of the project. Preparation of the quality plan is a major responsibility. It is normally entrusted to key members of the project team, headed by the project manager, and assisted by the quality manager in the head office. Input is also sought from other staff who are responsible for contracts, purchasing, logistics, etc. The draft quality plan should be reviewed by the people concerned before being adopted.

Quality planning, resulting in a well-conceived quality plan, is an important function of project quality management. The planning process should be covered by a documented procedure, an example of which is shown in section 9.3 on pages 158–66.

Quality planning preferably begins at the time when the tender is prepared. Submitting a quality plan with the tender, even in outline, is a demonstration to the client of your understanding of the project requirements and capability in assuring the quality of the finished works. (In a competitive market, this can be your competitive edge.) If not included in the tender, the quality plan is usually requested at the time of signing the contract, and is certainly needed before commencement of work. Where agreed in the contract, the quality plan has to be accepted by the client before being implemented.

## 3.2 QUALITY PLAN

A quality plan is a document setting out the specific quality activities and resources pertaining to a particular contract or project. Its contents are drawn from the company's quality system, the contract and related documents, as shown in Fig. 3.1. In the quality plan, the generic documented procedures are integrated with any necessary additional procedures peculiar to the project in order to attain specified quality objectives.

The quality plan is virtually a quality manual tailor-made for the project. The client, or the architect / engineer acting as his representative, may indicate in the contract what the quality plan must include and which items are subject to mutual agreement. For example, it is often specified that the inspection and test plans which invariably form part of the quality plan are to be approved by the architect / engineer prior to use.

Typically a quality plan contains most, if not all, of the following:

- brief description of project;
- list of contract documents and drawings;
- project quality objectives;
- site organization chart, with named personnel if known;
- responsibilities and authorities of project staff;

- site layout plan;
- construction programme and sub-programmes;
- schedules of subcontractor nomination, material and equipment procurement, based on the construction programme;
- list(s) of materials and appliances used for the project, showing the verification requirement of each;
- inspection and test plans, or list thereof;
- list of quality procedures and work instructions applicable to project - by making reference to the company's Quality Manual and Procedures;
- list of project-specific procedures, work instructions and inspection checklists, or target dates for their provision;
- list of quality records to be kept, including pertinent quality records from subcontractors;
- frequency (or provisional dates if possible) of internal quality audits;
- frequency of updating the quality plan.

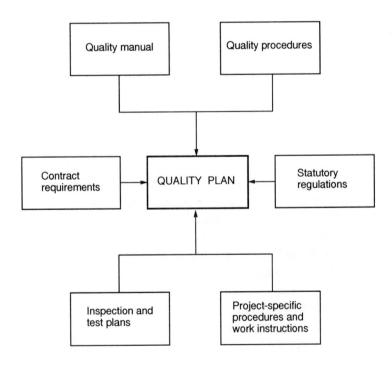

**Fig. 3.1** Contents of the quality plan.

A brief description of the project serves to clarify the scope of work undertaken by the contractor and to define the area of site within his responsibility. The site layout plan should show the access and passage for heavy vehicles, strategic location of lifting equipment and designated storage areas.

As project quality objectives, the contractor usually expresses his commitment to carry out the project in full compliance with the contractual and regulatory requirements, and to deliver the finished works on time. Consequently he should list the contract documents and drawings on which the requirements are defined. A construction programme, together with any sub-programmes, is included in the quality plan to show how he plans to achieve on-time delivery. This is supplemented with a schedule (or schedules) of procurement of labour, material and equipment necessary for the project.

The site organization chart should show all staff with a management or supervisory role. If someone has been assigned to a position when the quality plan is made, his/her name is attached to the position title. The responsibility and authority given to each position are briefly but clearly stated. There is normally a quality officer (or a similar title) among the site staff. While reporting directly to the project manager, the incumbent of this position has functional linkage with the quality manager in the head office.

Operations on site are regulated by the company's quality system, modified as necessary to suit the particular needs of the project. Invariably the quality plan calls up those quality procedures of the company that are applicable to the project. It is not necessary to include the procedures here but to make reference to the respective documents by number and title.[*] A few of the procedures may have to be modified to suit the particular contract, e.g. a longer-than-usual retention period of quality records or a specific time frame within which client complaints must be responded to. There are also work instructions and inspection checklists covering the construction and installation activities. If any of these documents does not already exist in the library of current work instructions etc., it is to be written by the project manager or someone under his instructions and the target date when it will be available is indicated in the quality plan.

Listing of materials and appliances to be used for the project is conveniently done by putting them into the following categories:

---

[*] Procedures referenced in the quality plan should be made available to the client or his representative on request.

- items that require approval of the architect / engineer before use;
- items that are supplied with the manufacturer's test certificates;
- items that require sample testing, indicating the type and frequency of test and, if appropriate, the testing laboratory commissioned.

While the inspection and testing needs of materials received on site are summarized in the lists mentioned above, those relating to works produced are covered by the inspection and test plans. The ITPs always form an essential part of the quality plan. Most of them can be retrieved from previous projects, with or without modification. If an ITP is not ready at the time when the quality plan is formulated, a target date should be set in the quality plan for its provision.

As construction proceeds, a volume of quality records is generated through implementation of the quality plan. The majority of quality records constitutes documentary evidence of product conformance. Typical records in this category are material receipt and dispatch forms, material test reports, weld inspection reports, piling records, ITPs duly signed off, and as-built drawings. Other quality records serve as verification that the planned activities of quality assurance have been taken. These records include inspection and test plans duly signed off, project-specific procedures and work instructions, and minutes of project management review meetings. The quality plan should show a list of all quality records to be collected and maintained.

Adherence to the quality plan is checked by internal quality auditing. Provisional dates of internal quality audits may be set, based on the anticipated progress of the project. If this is not possible, the frequency of audit is stated.

The quality plan is a 'live' document. It is updated periodically in the light of changes of contractual requirements, staffing and other circumstances which arise during the progress of the project. Normally the project manager reviews the quality plan, and amends it if necessary, at monthly intervals for an initial period of, say, six months, and thereafter when significant changes occur. An amendment sheet is included for recording the details of each amendment, such as the revision number, pages replaced, contents amended, date of effect and signatory evidence of updating. This usually forms the first page of the document.

The quality plan is a controlled document. Like other quality system documents, it is identified by the document number, the issue number, the revision number and the date of issue or revision. Issue and distribution of the document should follow the document control procedure of the quality system.

## 3.3 INSPECTION AND TEST PLAN

An inspection and test plan of a construction or installation process is a document which shows the verification measures (inspection and/or test) to be taken during the process and the criteria for acceptance of each inspection / test. It is normally presented in a table or flowchart, following the sequence of activities of the process in question.

Construction and installation processes for which ITPs are normally prepared include, but are not limited to, the following:

- foundation work, e.g. piling, caissoning and excavation;
- concrete work, including formwork and prestressing;
- precast work;
- erection of steelwork;
- waterproofing;
- plumbing and drainage work;
- installation of cladding, facade and curtain walls;
- installation of fire-fighting equipment;
- installation of building services;
- special processes such as woodwork for acoustic effects.

The inspection / test may be done by the supervisory staff on site or by an independent body such as a testing laboratory or a regulatory authority. Examples of the former type are: checking of dimensions and levels, inspection of reinforcement layout before concreting, etc.; examples of the latter type are: concrete sampling and testing, trial run of fire-fighting equipment, etc. Depending on the nature and importance of the work, the ITP is signed off either by a representative from the architect's / engineer's office, by the contractor's foreman, or just by the charge-hand of the gang who has finished the work.

An ITP is preferably prepared by the party carrying out that portion of work, either the contractor or the subcontractor. The architect / engineer normally reserves the right to approve the ITPs before use. As most of the processes are routine in nature, it is often possible to use the ITPs over and over again for different projects.

## 50    Project quality management

To prepare an ITP, break up the construction or installation process into distinct activities one after the other, and identify the inspection and/or testing required on completion of each activity. Then, for each inspection / test, define the following:

- person or party responsible for inspection / test (architect, engineer, contractor, subcontractor or testing laboratory);
- method of inspection / test, e.g. visual inspection, standard test, test procedure No. ....., checklist No. ....., etc.;
- acceptance / approval criteria, normally by referring to a certain clause in the specifications or a national standard;
- whether it is a witness point or a hold point;
- records to be retained.

A witness point is when the presence of an authorized person, such as someone from the architect's / engineer's office, is essential during the inspection or test. A hold point is when the approval of the authorized person is required before the work can proceed. The final inspection / test of a certain stage of work is always a hold point: the completed part of work is not allowed to be built upon or covered up until the authorized person is satisfied with its quality and releases the hold.

The architect / engineer, acting as the client's representative, can make use of witness points and hold points to monitor the verification activities of the contractor. It is here that the quality control efforts of both parties are integrated. The architect / engineer or somebody authorized by him will either witness the inspection or test (witness point), or in addition give permission for the work to proceed (hold point) if he is satisfied with the quality of the work so far. In a similar manner, the contractor, acting as client in a subcontract, uses the witness points and hold points in the subcontractor's ITP to monitor the services he receives. An example of an ITP is shown in Fig. 3.2.

ITPs are quality system documents, and as such are subject to document control. When duly signed off, they become permanent records demonstrating conformance to quality requirements.

### 3.4    PROJECT MANAGEMENT REVIEW

Quality management of a project, through implementation of the quality plan, is reviewed at regular intervals. With a building construction project,

**Quality Construction Limited**
**Building Contractors**

## INSPECTION AND TEST PLAN

Contract No. : 123/97  ITP No. : 5
Project : Sincere Insurance Building  Revision No. & date : 1 (10/12/97)
Subcontract : Plumbing  Work covered : Underground drainage pipes

| Activity | Inspection / Test | Acceptance criteria | Records / Remarks | Verifying party ||||  Sign & date |
|---|---|---|---|---|---|---|---|---|
| | | | | SC | C | A/E | RA | |
| Setting out | Check location | As per drawing A12 | | I | | | | |
| Trench excavation | Inspect base preparation | Spec. Clause 35.2 | Dewatering when req'd. | I | W | | | |
| Concrete bedding | Inspect concrete surface | Spec. Clause 35.2 | Grade 15 concrete | I | | | | |
| Pipe laying | Check alignment & fall | As per drawing A12 | | I | W | | | |
| Manhole / gully | Check invert level | Spec. Clause 35.3 | | I | W | | | |
| Smoke / water test | Check for leakage | No observable leakage | Test record | T | H | W | W | |
| Connect to govt. sewer | Inspect joint | No observable defect | As-built drawing | I | | | H | W |
| Backfill | Inspect compaction | Spec. Clause 35.4 | | I | | | | |

Verification code :
- I  Inspection
- T  Test
- W  Witness
- H  Hold
- D  Document review

Legend :
- SC  Subcontractor
- C  Contractor
- A/E  Architect/Engineer
- RA  Regulatory authority

**Fig. 3.2** Example of inspection and test plan.

such as a multi-storey building, a shopping centre or a block of apartments, an interval of four to six months is appropriate. However, the review has to be more frequent if noncomformance is encountered time and again.

Project management review is usually by way of a meeting of the people concerned. The review meeting is convened by the project manager, assisted by the quality officer on site. Naturally the site agent, the general foreman and the engineer would be involved. Certain sub-contractors may be invited to attend, especially when nonconformities are found in their work.

In the review meeting, the impact of changes in contractual requirements or redeployment of staff and other resources is considered. Nonconformities discovered and client complaints received since the last review are brought up for discussion. Effectiveness of any corrective or preventive action taken is assessed. In addition, training needs arising from staff changes or quality deficiencies are identified. The performance of subcontractors and suppliers is evaluated at the same time.

The review meeting is recorded. A copy of the minutes is forwarded to the quality manager who will place it on the agenda of the corporate management review meeting when it is next held.

The quality plan is updated to incorporate changes made. The revised document is issued and distributed under control.

## 3.5  CLIENT'S ROLE IN QUALITY ASSURANCE

Quality assurance starts with the client. In order to get the desired quality, i.e. fitness for purpose, the client has to define the purpose of the product in the first instance. In other words, the architect / engineer, acting as his representative, has to express the requirements unambiguously, explicitly and clearly in the contract drawings and specifications.

In a traditional contract, the architect / engineer serves as a supervisor external to the contract, and both parties of the contract tend to accept the judgment of this supervisor as a yardstick of acceptable quality. In a quality assurance based contract, the onus of supervision and verification lies with the contractor, although the architect / engineer usually reserves the right to monitor the verification work through the use of hold points and witness points in the inspection and test plans. Consequently the specifications must be capable of interpretation and application without reference to an external supervisor (Barber, 1992). Phrases like 'to the satisfaction of the architect' which do not clearly show the desired quality would have no place in the contract. The specifications should also identify the hold points and witness points, and indicate how much prior notice is required.

In addition to the above, the client or his representative should take the following actions in a well-considered, timely manner:

(a) Selects the appropriate quality system requirements for each contract.
(b) Clearly specifies the quality system requirements in tender and contract documents.
(c) Evaluates and selects subcontractors on their ability to satisfy specified requirements.
(d) Reviews and accepts the quality system documentation specified for the supplier.
(e) Monitors the works and the implementation of the quality system.
(f) Collects, reviews and controls the quality records that the supplier is contracted to provide (SA/SNZ, 1997).

In a quality assurance based contract, the client should indicate in the contract whether the contractor's quality system is to comply fully or partly with ISO 9001 / ISO 9002. (If in parts, which quality system elements are applicable to the project.) The client normally also specifies the various quality-related documents the contractor has to provide and the time of submission of each document. Table 3.1 shows a matrix of when the various documents are to be made available (SA/SNZ, 1997). The major document that the client would ask for is the project quality plan. Normally this is for information only, but the architect / engineer may specify that the quality plan be subject to his approval. Before putting such requirement in the contract, be aware of its implication: involvement in the evolution of the document carries certain responsibility and any responsibility has the potential for liability. In any case, it would be prudent to avoid such definitive terms as 'approve' or 'disapprove' when commenting on the contractor's planned quality activities.

An organization engaged in construction often acts as a 'purchaser' as well as a 'supplier'. For instance, the contractor is a supplier of construction services to the client and at the same time a purchaser of subcontracted work. Similarly, a subcontractor, in providing certain services to the contractor, is being provided with raw materials by various manufacturers and importers. The network of purchaser - supplier interaction is illustrated

*54 Project quality management*

in Fig. 3.3 (SA/SNZ, 1997). The complex interaction, in fact, is a series of two-party contractual relationships to which ISO 9001 / ISO 9002 applies.

When the contractor acquires the services of a subcontractor, he has the same responsibilities as a client. He must prepare the subcontract in such a way that it clearly and truly reflects the requirements specified in the main contract. He also has to indicate at what stage of the subcontracted work his site agent or general foreman will be present for the inspection or test.

## 3.6 SUMMARY

- Quality planning of a project includes identification of necessary resources, training requirements, inspection and testing needs, and quality procedures applicable to the project. It should be carried out before construction work is due to commence. The outcome is summarized in a quality plan.
- A quality plan is virtually a quality manual tailor-made for a project. Its contents are drawn from the company's quality system, the contract and related documents.
- An inspection and test plan enables a systematic approach to carry out in-process verification of work. It lays down in sequence the inspection and testing involved in a construction or installation process, and for each inspection or test the person who has the authority of approval. When duly signed off, it provides documentary evidence that the quality of work has been verified.
- Project quality management, through implementation of the quality plan, is reviewed at regular intervals.
- In the spirit of cooperation, the client has the obligation of stipulating the project requirements unequivocally in the contract documents, including requirements pertaining to quality assurance.

**Table 3.1** Documentation submission matrix (SA/SNZ, 1997)
*(reproduced by permission of Standards Australia / Standards New Zealand)*

| Document description | With tender | | | During contract | | |
|---|---|---|---|---|---|---|
| | Copy to be submitted | Sample to be submitted | To be submitted for review ..... days prior to use | To be available for review on request | | To be submitted on completion |
| Company quality manual | ✓ | o | o | + | | o |
| Quality procedures | o | o | + | ✓ | | o |
| Project quality plan | o | + | ✓ | + | | o |
| Design plan | o | + | ✓ | ✓ | | o |
| Inspection and test plans | o | + | ✓ | ✓ | | o |
| Technical procedures | o | o | ✓ | ✓ | | o |
| Quality records | o | + | + | ✓ | | ✓ |
| Other quality requirements | + | + | + | + | | + |

LEGEND :  ✓  usually required
+  may be applicable
o  not usually applicable

NOTE :  The number of copies will need to be indicated in the relevant boxes.

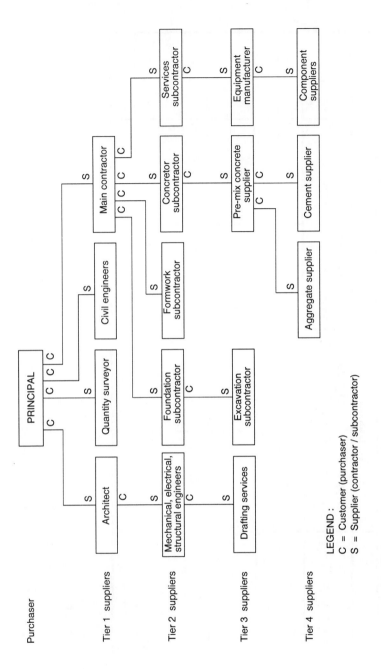

**Fig. 3.3** An example of purchaser-supplier network in the construction industry (SA/SNZ, 1997). *(reproduced by permission of Standards Australia / Standards New Zealand)*

# 4

# Developing a quality system

## 4.1  WHERE TO START

Quality assurance of construction works begins at the time the project brief is developed and passed to the architect / engineer. The client, whether a government body or a private enterprise, operates a quality system to ensure that the requirements of the project are adequately defined and clearly communicated to the architect. The architect / engineer makes use of another quality system to control the design process such that the client's requirements are fully and unequivocally incorporated in the drawings and specifications. In turn, the contractor implements yet another quality system to ensure that the construction is carried out in accordance with the specified details. In this book, attention is focused on helping the contractor establish a quality system for his organization. It should not be forgotten, however, that quality of the final product can only be assured with the concerted efforts of all parties involved.

Every construction company has its own business policy and traditions. Its workforce may range from a few persons to dozens of people. It is utmost important to establish a quality system that is *suitable* for the organization, depending on the scope of its operations and the size of its workforce. The quality system should not be a managerial strait jacket; instead, it should resemble a tailor-made coat - tight-fitting yet providing room for manoeuvre.

The company would have some sort of quality system in place, otherwise it could not survive in the competitive market. There may be a number of *standard procedures* already in use. However, the list of procedures is probably not comprehensive. Some activities are rationalized while others are not under proper control. Besides, some established procedures are occasionally ignored or not strictly followed for a variety of reasons. Above all, these procedures are not fully documented and regularly audited, resulting in inefficiency and even confusion.

In most cases, developing a quality system does not start from scratch. The task ahead is to remould the existing sketchy system into an integrated system along the lines of ISO 9001 / ISO 9002. Either standard specifies a full set of requirements of a quality system, but leaves it to the individual organization to decide which elements its quality system has to encompass. The requirements of both standards are the same, except that ISO 9001 includes design control while ISO 9002 does not. In building construction, it is traditional practice that a contract is let on the basis of a design produced by an architect, and the contractor does not design any permanent works. If that is so with your company, you may develop your quality system conforming to ISO 9002. (Design of temporary works such as shoring and formwork may be considered as part of the construction process and therefore covered by process control.) If your company undertakes design-and-construct contracts as well, you should go for ISO 9001.

Following an international standard right from the start has several advantages. Firstly, it ensures that no important activities of quality assurance are missed out. Secondly, it unveils the quality deficiencies of the current practices when compared to the preferred practices. Thirdly, it paves the way for third party certification (or registration) which should be the corporate goal.

Establishing a quality system is an important endeavour of the company in its evolution, with repercussions on operational efficiency and business prospect in the years to come. Senior management should be aware of its significance and show full commitment. Failure to provide adequate resources for the task and reluctance to release staff for quality awareness training are obvious signs of lack of commitment. It is worth remembering that a half-hearted approach is bound to fail, creating more paperwork but not improving the output.

Despite its importance, establishing a quality system is no big deal. This simply means putting the house in order: clarifying everyone's responsibility, putting current practices in writing and ensuring that they are adhered to at all times. In fact, a quality system is just 'common sense set down in a structured way'. To give an example, it is common practice for the general foreman to inspect the building materials received on site, and a quality system just requires him to keep a record each time this is done.

Before embarking on the endeavour, it is worthwhile to learn from your predecessors. Try to gather (and analyse) the hard-earned experience of those who have done it, some of whom may be your competitors. In fact, the collective wisdom worldwide has been distilled into five unfailing guidelines as follows:

- Involve as many staff as possible in the development process.
- Establish the quality system to suit the organization, but not the organization to suit the quality system.
- Develop the quality system at a pace that the organization can cope with.
- Keep the documentation to the minimum, yet enough to achieve effective control.
- Implement the quality system progressively.

## 4.2 GETTING STARTED

To get started, a person is appointed to be in charge of this specific task. A senior member of staff who has years of service with the company and enjoys a high degree of respect at all levels would be most suitable. This person must have access to, and the confidence of, the top management. Under his/her leadership, a quality assurance team or *taskforce* is set up. Members of the team should include people from management as well as representatives of the workers.

The first job of the QA taskforce is to appraise themselves of their knowledge of quality assurance principles and practice. The outcome of the appraisal will lead to the decision whether to employ a quality consultant. The service offered by a quality consultant ranges from an induction course for the QA taskforce to a complete package of documentation, training, implementation and fine-tuning of the quality system. Generally speaking, the more complex the company's operations are, the greater is the need of the consultant's service. A word of warning: a certification body (or any individual staff of the body) whose service the company intends to use later cannot be employed as a consultant because of obvious conflict of interest. Nevertheless, the certification body may be approached for advice on the adequacy of the quality system at the later stage of its development.

While a consultant is very helpful in kick-starting the process of system development, the need of his/her service diminishes as the development work proceeds. In fact, the consultant often lacks the knowledge of the company's operations to continue to make essential contribution after the development work is put on its track. The best way is for the consultant to

## 60 Developing a quality system

train up the members of the QA taskforce and guide them in developing the quality system by themselves.

The process of establishing a quality system (including third party certification) takes about fifteen to thirty months. It involves a lot of staff at various stages and consequently must be properly managed right from the beginning. Necessary actions are identified and put into a logical sequence. These normally include, but are not limited to, the following:

- reviewing current practices;
- preparing documentation;
- training operators in documented procedures;
- implementing the quality system;
- internal quality audits;
- reviewing the quality system;
- arranging for third-party certification;
- pre-assessment audit;
- certification audit.

It is advisable to draft an action plan in the form of a bar-chart, such as the one shown in Fig. 4.1. By necessity some activities overlap in time. Prepare also a budget at the start so that the top management knows how much to provide for as the work proceeds.

### 4.3 REVIEWING CURRENT PRACTICES

The review process is, in essence, to scrutinize the written or unwritten procedures and to rationalize them in the process. This is achieved by conducting interviews, observing how the operators work and examining whatever documentary evidence is available. It is advisable to prepare some worksheets beforehand for collecting information in the interview. They are helpful in eliciting responses. They also ensure that the gaps in the quality system are closed. A worksheet for reviewing the activity of 'evaluation of subcontractor' is shown in Fig. 4.2. As you can see from the example, the worksheet contains simple questions based on your interpretation of the requirements of the International Standard when applied to your company. A complete set of worksheets may be found in (Novack, 1994). Although they are designed mainly for the manufacturing industry, these worksheets show the way for you to make up your own.

| Activity | 1 | 2 | 3 | 4 | 5 | 6 | 7 | 8 | 9 | 10 | 11 | 12 | 13 | 14 | 15 | 16 | 17 | 18 | 19 | 20 | 21 | 22 | 23 | 24 |
|---|---|---|---|---|---|---|---|---|---|---|---|---|---|---|---|---|---|---|---|---|---|---|---|---|
| Appoint QA Manager | ■ | | | | | | | | | | | | | | | | | | | | | | | |
| Establish QA team | | ■ | | | | | | | | | | | | | | | | | | | | | | |
| Prepare and approve budget | | | ■ | | | | | | | | | | | | | | | | | | | | | |
| Appoint QA consultant (optional) | | | ■ | ■ | | | | | | | | | | | | | | | | | | | | |
| Brief QA team on QA concepts | | | | ■ | | | | | | | | | | | | | | | | | | | | |
| Develop quality policy | | | | | ■ | | | | | | | | | | | | | | | | | | | |
| Publicize QA concepts among staff | | | | | | ■ | ■ | ■ | | | | | | | | | | | | | | | | |
| Review current practices | | | | | | | | ■ | ■ | ■ | | | | | | | | | | | | | | |
| Prepare quality system documents | | | | | | | | | | | ■ | ■ | | | | | | | | | | | | |
| Train staff in use of quality procedures | | | | | | | | | | | | | ■ | ■ | | | | | | | | | | |
| Implement quality system | | | | | | | | | | | | | | | ■ | ■ | ■ | | | | | | | |
| Train internal quality auditors | | | | | | | | | | | | | | | ■ | | | | | | | | | |
| Select and approach certification body | | | | | | | | | | | | | | | ■ | | | | | | | | | |
| Submit quality documents to cert. body | | | | | | | | | | | | | | | | ■ | | | | | | | | |
| Conduct internal audits | | | | | | | | | | | | | | | | | ■ | ■ | | | | | | |
| Arrange pre-audit visit | | | | | | | | | | | | | | | | | | ■ | ■ | | | | | |
| Revise quality system, if necessary | | | | | | | | | | | | | | | | | | ■ | | | | | | |
| Prepare for system audit | | | | | | | | | | | | | | | | | | | | ■ | | | | |
| Correct nonconformance, if necessary | | | | | | | | | | | | | | | | | | | | | | ■ | ■ | |
| Prepare for follow-up audit, if necessary | | | | | | | | | | | | | | | | | | | | | | | | ■ |

Month

**Fig. 4.1** Typical action plan for quality system development and certification.

## Developing a quality system

---

**Worksheet for Review of 'Evaluation of subcontractors'**
Reference : ISO 9002 Clause 4.6.2

1. What types of jobs are subcontracted?
   [ ] material supply    [ ] special processes
   [ ] labour supply      [ ] design (e.g. design of shoring)
   [ ] plant hire         [ ] equipment maintenance &calibration
   [ ] building services  [ ] others

2. Are all subcontractors evaluated before employment?
   [ ] Yes                [ ] No
   If no, under what conditions is an exception made?

3. Who is in charge of evaluation of subcontractors?

4. What are the measures for evaluation?
   [ ] labour resources   [ ] quality system
   [ ] plant resources    [ ] financial standing
   [ ] workmanship        [ ] track record
   [ ] on-time delivery   [ ] references
   [ ] health & safety    [ ] others

5. Is there a list of acceptable subcontractors maintained for each trade?
   [ ] Yes                [ ] No
   If yes, who keeps and updates the list?

6. Are there restrictions set for each subcontractor on the list?
   [ ] Yes [ ] No
   If yes, what are they?
   [ ] contract sum       [ ] complexity of work
   [ ] volume of work     [ ] others

7. Are subcontractors always selected from the appropriate list?
   [ ] Yes                [ ] No
   If no, under what conditions is an exception allowed?

8. Are subcontractors on the list evaluated at regular intervals?
   [ ] Yes                [ ] No
   If yes, how frequently is this done?

9. If a subcontractor on the list does not measure up to expectation, what are the steps to take to delete it from the list and by whom?

---

**Fig. 4.2** Typical worksheet for evaluation of existing practices.

During the review, weaknesses of the current practices are exposed, ambiguities are clarified and contradictions removed. Other than these, it is not advisable to make changes to the current practices unless the modifications are necessary to satisfy the requirements of the International Standard. To facilitate a *smooth ride* when implementing the system, the procedures should be documented more or less as they are.

As the operators know their job best, they should be involved in the review process under the direction and coordination of the QA taskforce. Participation at the development stage promotes a feeling of contribution and brings about willing cooperation when the system is implemented. These people may be asked to write down step-by-step how they carry out their work. The information so collected forms the basis on which proper procedures are prepared later.

In developing a quality system conforming to ISO 9001 or ISO 9002, or modifying an existing quality system to the same end, the current practices are brought in line with the 'quality system requirements' of the International Standard. The clauses specifying the requirements, twenty altogether, are addressed one by one. Each clause is paraphrased in the context of the company's operations. In doing so, you may find that some existing procedures need to be revised and some new procedures need to be established. On the other hand, a few clauses such as *4.4 Design control* and *4.20 Statistical techniques* may not be applicable to your organization. In deciding whether a standard procedure is required for an activity, you should take into consideration the complexity of the work, the method used, and the skills and training needed by personnel carrying out the activity. Moreover, you should judge by past history whether the absence of such a procedure would substantially increase the risk of unsatisfactory outcome.

## 4.4 PREPARING QUALITY SYSTEM DOCUMENTS

An essential requirement of a quality system is that it is fully documented. The key document is the quality manual.

A quality manual is a 'document stating the quality policy and describing the quality system of an organization' (ISO, 1994a). It gives an overview of the quality system and outlines the structure of the system documentation. It includes or makes reference to the quality procedures that form a major part of the system documentation. In simple terms, the quality manual lays down what the company intends to do to achieve quality.

Internally the quality manual serves as the bible for carrying out the company's operations. Externally it helps build up the client's confidence in the company's capability for, quality work. It is also the basis for quality audits and certification.

The quality manual may be a single-volume document covering everything pertaining to the quality system, but more often it is one document supported by other documents in several tiers, each tier becoming progressively more detailed. The 'document pyramid' in Fig. 2.2 (page 13) illustrates the hierarchy of documentation as quality manual, quality procedures and work instructions. The quality manual shows the framework of the quality system while the quality procedures describe the administrative processes and the work instructions (made under the procedures) give the working or technical details. Separating the quality procedures from the quality manual enables the company to restrict the trade secrets to its own staff while releasing non-confidential information to potential clients.

It is arguable whether the quality manual or the quality procedures should be written first. Starting with the quality manual has the advantage of fully declaring the company's intentions before establishing the methods to achieve them. On the other hand, writing the quality procedures first tends to make them more reflective of the company's current practices. In the author's view, the two sets of documents should be developed hand in hand. In fact, the manual resembles the structural frame of a building and the procedures the prefabricated units. While the structural frame is being erected on site, the units are produced in a precast factory. Drafting of procedures proceeds parallel with review of current practices and it will take quite some time to complete. Meanwhile, preparation of the quality manual may go ahead, incorporating the written procedures as they become available and bringing to light the new procedures necessary. By doing so, incompatibilities are expediently detected and remedied.

Quality system documents are controlled documents. Each document should be reviewed and approved before issue. For control of distribution, any controlled copy should have the copy number and the name of the receiver inscribed conspicuously somewhere on the document such as the title page.

Quality system documents are meant to be consulted by staff at all levels. They should be concise but precise, and at the same time easily understood. Guidelines for document writing are given in Chapter 8 and sample documents are provided in Chapter 9.

## 4.5  QUALITY RELATED TRAINING

To make a quality system work, it is essential that all personnel understand what it is about and what role each has to play. However, the majority of staff are not yet familiar with the concepts of quality assurance and its implications. Consequently the importance of quality related training cannot be over-emphasized.

Appropriate training should be provided to different levels of staff at different stages of development and implementation of the quality system. At the initial stage when the quality policy is formulated, members of the QA taskforce often benefit from a briefing or teach-in by a quality consultant. They will in due course carry out a quality awareness campaign among their colleagues in the respective sections of the organization.

Another type of training is also needed for those who are to draft the quality documents. Again the quality consultant may be called upon to provide the training. The intensity of training depends on the writing skill of the drafting team.

At a later stage, a series of workshops or simulation exercises should be arranged, both in the office and on site, to familiarize the operators with the documented procedures that they are to follow in their daily work. The QA taskforce would be able to act as instructors as they have been involved in the development work. In fact, this provides an opportunity for them to test-run the quality system. It would be effective to separate the staff into different groups and to emphasize on the aspects that are of direct concern to the area of operation of a particular group. Regular subcontractors and their supervisory staff may be invited to join in the training programme.

Another area where training is essential is quality auditing. Staff from one section are selected to audit the activities in another section. They should be equipped with the basic techniques of auditing in order to do a proper job. A quality system will not bring about the expected results if the internal quality auditors are unable to discover the quality deficiencies or disinclined to disclose them in their report.

## 4.6 SUMMARY

- To establish a quality system, a senior member of staff is appointed to be in charge, assisted by a taskforce drawn from management as well as the workforce.
- To establish a quality system and get it certified by a third party takes about fifteen to thirty months. An action plan should be prepared, preferably in the form of a bar-chart, indicating the time sequence of various actions.
- Current practices in the company are reviewed, and modified if necessary, to meet the requirements of the International Standard. In this respect, the operators can make a significant contribution because they know their job best.
- The quality system is documented in a quality manual and a set of quality procedures. The quality manual sets out the intended quality functions while the quality procedures provide details for action. Together they address and satisfy all applicable requirements of the International Standard to which the present quality system is to conform.
- Training is an essential element for success. At the initial stage of development of the quality system, training is mainly concerned with awareness of the quality assurance principles and requirements. At a later stage, training takes the form of simulation exercises so as to familiarize the operators with the documented procedures that they are to follow in their daily work. Training should also be provided for staff who are to act as internal quality auditors.

# 5

# Implementing a quality system

## 5.1  THE TRIAL PERIOD

When the major parts of documentation are completed, it is time to put the quality system into practice. Translating the documented procedures into smoothly operating functions requires time and patience. The earlier this is started, the more time will be available for all concerned to understand and accept the established routine.

Implementing a quality system normally takes three to six months. You may introduce the procedures one by one as they are documented. People-on-the-job should be encouraged to identify and report any difficulties encountered. Monitor the progress through internal quality audits as described later in this Chapter. With the experience collected, you can fine-tune the quality system and revise the relevant documents.

No matter how carefully and intelligently the quality system documents have been prepared, they are unlikely to be without a hitch. Each procedure has to be tested just like a computer programme being debugged. Give each procedure a test-run as soon as it is written. Sometimes the procedure needs changing because it is found to be impractical or inefficient. At other times, firmness and perseverance are necessary, where the text correctly specifies the appropriate action.

A quality system is only a means to an end: it relies on *people* to make it work. In an organization, the performance of an individual could directly or indirectly affect the quality of the product. Quality assurance necessitates the concerted effort of all concerned. Many a quality system breaks down just because a few people do not adapt themselves to it. Staff at all levels must be willing to work to the established procedures and know how to do it. To promote willingness, they have to be motivated. To bring about awareness, they have to be trained. Motivation and training should go hand in hand, and together they constitute the recipe for success.

## 5.2 MOTIVATION

Instilling a new quality system into an organization produces a cultural change, even though the system has been developed from existing practices. Inevitably different reactions are generated among the staff. Some people jump on it as a challenging endeavour. Others are bewildered and go along with reluctance. Yet others become openly resistant to the change and persist in following their long-held habits. Those in middle management are especially sceptical of the necessity for change as they cannot find tangible benefits right away.

It should be remembered that quality assurance is the joint responsibility of management and the workforce. Everyone in the organization must realize the importance of quality output to the company and be willing to play his/her part in achieving it. Without the genuine support of all concerned, the benefits of quality assurance will never materialize.

The majority of staff, including some in managerial positions, need to be motivated to adapt to the cultural change. This is no easy task for a building construction company that is made up of a group of people with grossly different background, education and aspirations. Paying a bonus may work, but there will be negative effects when the monetary incentive is stopped. Penalizing the unconvinced is counter productive, for this will make them more resistant to the change. Instead, the following motivators have been proven to be effective:

- commitment and leadership;
- staff involvement;
- training opportunities;
- recognition of accomplishment.

Throughout the debugging stage of the quality system, lots of teething problems crop up and a certain amount of resentment is unavoidable. Even those who have helped develop the quality system tend to lose faith in it. At this crucial moment, it is up to senior management to show firm commitment and leadership in steering the ship through the stormy sea. They must demonstrate by action their willingness to face the problems and to allocate resources for their speedy resolution. Section managers of the organization must be convinced that this is the way to go and be seen practising what they preach. There is no better way to motivate others than setting an example yourself. Furthermore, the responsibility for quality must filter down the line of management. Quality assurance will not be achieved unless middle management feel that they are also in the driver's seat.

When the formalized procedures are first introduced, the paperwork involved tends to slow down production. This should not be construed as inefficiency of the quality system or incompetence of the staff. (The situation will improve as people become more adapted to the system, otherwise the system is not suitable for the organization.) Instead of penalizing the staff for reduced productivity which is temporary, the managers should provide additional resources to ease the pressure of time.

Some people in the organization may mistakenly think that a predetermined system is thrust upon them. Never should this feeling be allowed to take root. Indeed, a quality system cannot be forced onto the unwilling. It is worth remembering that quality is the product of the coordinated efforts of the entire organization, not the work of a few people or a particular section. That is why a dialogue must take place between the QA taskforce and the operators right from the start. In formalizing the procedures, consult the people who perform the various functions. Let them help in drafting the procedures, or keep them abreast of the development if they are not so involved. During implementation, provide opportunities for comment and treat the feedback seriously. All in all, this will promote a sense of involvement in hammering out a *workable* system, and will earn the support of the unconvinced and the reluctant. Most people are motivated by the satisfaction of making a meaningful contribution.

Some staff may query the wisdom of imposing the apparently bureaucratic measures. They usually argue that filling in the standard forms has turned people into slaves of paperwork. Haunted by this attitude, they just play lip service to the quality system. To win them over, you have to increase their awareness of the quality system. This may be achieved with appropriate training. When they are familiar with the system, they will realize that the standard forms actually serve as a reminder (and a record) of what has to be done and as such can only assist the construction process to run smoothly.

It is a natural inclination of people to desire that their efforts be duly recognized. They will be motivated if they can see that rewards are linked to efforts. As an incentive for quality work, the company may make it known to the staff that evidence of faithful support of the quality system is a criterion for promotion and, on the contrary, repeated neglect of the established procedures is a cause for disciplinary action. Combined use of the carrot and the stick often gets things done.

## 5.3 TRAINING

A quality system, by itself, is no more than a sophisticated skeleton. No matter how well it is articulated, it requires human intelligence to bring it to life.

The above analogy points out the importance of people in putting the quality system to work. The potential of the quality system cannot be exploited until the staff fully understand how it functions.

To promote awareness of the quality system, a well-planned and timely training scheme is essential. The responsibility for quality training lies initially with the QA taskforce that has developed the quality system.

Quality training, as distinct from training for specific skills, is carried out in two stages. While the quality system is being developed, the concepts of quality assurance are preached among the staff at all levels. It is particularly important to ensure that every member of staff understands the company's quality policy and realizes the management's firm commitment to quality. At an appropriate time, a formal function may be held for the whole company at which the chief executive launches the company's quality policy. This is followed by informal meetings at various levels. Members of the QA taskforce spread the gospel in the various sections of the company under their command. The main purpose of these gatherings is to mould the attitudes of all personnel towards the cultural change in the organization. Leaflets may be prepared, describing those aspects of the quality system affecting a particular group. The contents of these leaflets should be adjusted to the level of understanding of the recipients. All the time, the quality policy is conspicuously displayed in places such as staff notice boards and the reception area. This serves as a constant reminder to the staff of their responsibility for quality assurance.

When the quality procedures are ready, it is time to embark on the second stage of quality training. Workshops are organized for the relevant groups of staff to familiarize them with the procedures, including the correct way of filling in the standard forms. Mock-up exercises are convened both in the office and on site. The construction industry is notorious for incompetence in written communication: the trade foremen may be proficient in carrying out supervision and inspection, but they often have difficulty in completing the various forms involved. The staff definitely need some practice before feeling comfortable with the new routine.

Quality training in an organization is an ongoing process. It is required not only when the quality system is first implemented but also after the system is in full operation. New staff are indoctrinated soon after joining the company. Existing staff are also given a refresher course every now and then, especially after major changes have been made to the procedures.

It is advisable to document the procedure of quality training so that it is conducted regularly and consistently. An example of such procedure is shown on page 241.

## 5.4 INTERNAL QUALITY AUDITS

Functioning of the quality system is monitored through a regular programme of internal quality audits coupled with periodic management reviews. An internal quality audit is a systematic and independent examination, performed by the company's own staff, to verify whether quality activities and related results comply with planned arrangements and whether these arrangements are suitable to achieve the quality objectives. It should not be confused with 'inspection' or 'surveillance' that is performed for the purpose of process control or product acceptance.

A series of internal quality audits should start as soon as possible, say, two or three months after the documented procedures are in use. The audits tend to expose the teething problems of the quality system so that they can be speedily resolved. The audits also evaluate the quality system in practice: its strengths and weaknesses as well as its acceptance by the staff. In other words, internal quality auditing leads to continuous refinement of the quality system.

Objective evaluations of quality system activities by competent personnel should include the following activities or areas:

(a) organizational structures;
(b) administrative, operational and quality system procedures;
(c) personnel, equipment and material resources;
(d) work areas, operations and processes;
(e) products being produced (to establish degree of conformance to standards and specifications);
(f) documentation, reports, record-keeping (ISO, 1994b).

Internal quality auditing provides evidence on how much the staff understand and follow the quality procedures; it also gives an idea of how well the quality system fits in the operations of the company. Through the auditing process, deficiencies of, and deviations from, the documented procedures are revealed. It is often found that certain procedures are not strictly adhered to, and some other procedures are impractical or difficult to interpret. Departure from a documented procedure noted in the audit provides an opportunity to streamline the process - either by changing the way of working or by educating the staff in the way the work should be done. Through corrective action that follows, the quality system is improved. It should be noted, however, that an audit of any kind deals only with a sample: absence of irregularities in the sample does not necessarily mean that they do not exist, but if they are found they would probably occur again unless appropriate action is taken to prevent their recurrence.

Internal quality auditing is itself a quality activity covered by documented procedure(s). It is subject to evaluation as other quality activities, i.e. the auditing process must be audited.

Internal quality auditing can be rather sensitive if not managed properly. The auditees may feel that it is a check by management on their performance. To ease their nervousness, management should assure the staff that the audit is carried out to evaluate the quality system but not the individuals. Furthermore, the audit is performed to establish facts rather than faults. After they get used to it, the staff will welcome the audit as a stepping stone to improvement. Sincerity and cooperation are central to an effective audit.

An internal audit is a planned exercise. The functional unit or construction site being audited is given ample notice of the intended visit. Not only should the auditors prepare for the task, but also the auditees must get themselves ready for interview and the quality records ready for inspection.

### 5.4.1 Quality auditors

Quality auditing is a highly professional activity. It will produce meaningful outcome only when performed by people with the right temperament and experience. Attributes of a good auditor appear to be a combination of analytical skill, communication skill, impartiality, diplomacy, keen observation and sound judgement. He/she is expected to be familiar with the quality system and always ready to listen rather than to express his/her own views.

Quality auditing promotes quality awareness and brings good learning experience. Anyone irrespective of rank and interested in the task should be encouraged to participate. A pool of internal quality auditors drawn from different levels of staff should be maintained. They are a mixture of managers, supervisors and general staff. About 10-20% of the entire staff would be involved.

The internal quality auditors, being either technical or administrative personnel of the company, carry out quality auditing as a secondary function on an ad-hoc basis. They are unlikely to possess the skill and experience to perform the task efficiently. (Even the quality manager may not be suitably qualified and experienced in this respect.) Consequently, training of internal quality auditors is essential. The staff concerned are usually sent to attend a short course (typically 2 or 3 days) offered by a professional body. In addition, in-house training may be given by the more experienced. Mock-up exercises would be enlightening to both the auditors and the auditees. These should take place before the quality system is in full operation.

### 5.4.2 Planning a quality audit

In a construction company, internal quality auditing goes on two fronts: system audits and project audits. A system audit evaluates the application of the quality system across the board. A project audit looks at the application of the quality system in a specific project and in particular the implementation of the project quality plan. In both cases, it is essentially to verify that documented procedures are available, understood and followed.

Overall planning of internal quality audits is the responsibility of the quality manager. Audits are scheduled as a routine exercise, as follow-up to a corrective action and after significant changes are made to the quality system or processes. Some activities are evaluated more frequently while others are subject to occasional checks. A series of audits is programmed such that each element of the quality system and each section of the company is evaluated at least once in a 12-month cycle. Audits of a project are usually scheduled at four to six months intervals and are conducted mainly on site.

In setting the frequency of audit of a particular activity, consideration is given to the importance of the activity and the results of previous audits.

An activity should be audited more often if one or a combination of the following is true:

- the activity is a critical link in the administration or construction operation;
- the activity is a constant source of nonconforming work or client complaint;
- a major nonconformity was noted in the last audit;
- a number of minor nonconformities were noted in the last audit;
- the operators are not experienced in the activity or use of equipment.

In preparing the master programme of internal quality audits, the quality manager has to decide whether to audit the company section by section or the quality system element by element. Auditing by location is simpler to administer. It is most suitable for project audits because all the quality system elements applicable to the project are evaluated in one exercise, thereby avoiding the necessity of multiple visits to the same construction site. A typical programme is shown in Fig. 5.1. Auditing by location, however, has a major drawback: the auditors tend to concentrate on the quality element(s) pertinent to the mainstream activities at the location and may overlook the other elements which also apply there. Auditing the quality system element by element, on the other hand, ensures that the entire quality system will be evaluated in one cycle. The auditors are also better prepared as they need to deal with only a couple of procedures at a time. Nevertheless the quality system element being audited, e.g. document control or corrective action, may apply to several sections of the company, and the auditors must visit all these sections to complete the audit trail. In any case, the master programme is provisional only. The frequency of future audits of a particular activity has to be altered if unexpected results, either good or bad, are obtained in the current audit. The quality manager will exercise his discretion in revising the audit programme to suit the needs that arise.

Detailed planning of an internal quality audit starts with the selection of auditors from the group of staff who have been so trained. It is usual to assign two (or more) auditors to conduct an audit, unless the audit is simple enough to be handled comfortably by one person. A more qualified and experienced person is named as the lead auditor who will head the audit team. Preferably he/she is one who has acted as an auditor for at least

**Quality Construction Ltd.**
**Building Contractors**     Internal Quality Audit Programme for 1998

| Section / Site | Scope of audit | Jan | Feb | Mar | Apr | May | Jun | Jul | Aug | Sep | Oct | Nov | Dec |
|---|---|---|---|---|---|---|---|---|---|---|---|---|---|
| Administration | Management review | | | | | | ◪ | | | | | | |
| | Training | | | | | | | | | | | ◪ | |
| Contracts | Tender and contract review | | | | | ◪ | | | | | | | |
| | Purchasing and subcontracting | | | | | | | ◪ | | | | | |
| | Control of contract documents | | | | | | | | | ◪ | | | |
| Construction | Quality planning | ◪ | | | | | | | | | | | |
| | Control of nonconforming work and client complaint | | | ◪ | | | | | | | | | |
| | Corrective and preventive action | | | | | | ◪ | | | | | | |
| Quality assurance | Control of quality documents | | | ◪ | | | | | | | | | |
| | Control of quality records | | | | | | | ◪ | | | | | |
| | Internal quality audits | | | | | | | | | | | ◪ | |
| Site 1 | Quality plan 1 | ◪ | | | | | | | | | | | |
| Site 2 | Quality plan 2 | | | | | | | | | | ◪ | | |
| Site 3 | Quality plan 3 | | | | | | | ◪ | | | | | |

Audit No. ◪ —— Audit date

Prepared by : *R.S Lee*

Date : 22/12/97

**Fig. 5.1** Typical programme of internal quality audits.

three audits. (The company will build up the pool of experienced auditors with time.) The auditors must be functionally independent of those responsible for the activity being audited. This is necessary to preserve the objective view of the auditors. However, technical staff from one site may be called upon to carry out an audit on another site with which they have no connection. It is not advisable to send the same people to the same place every time, for they tend to concentrate on the same aspects. A fresh team often has new discoveries.

An internal quality audit may last from a few hours to a few days depending on the complexity of the activity (or activities) to be audited. A time schedule of the audit is worked out showing the activities to be evaluated, the persons expected to be present and the time allocated. The schedule is delivered well in advance to the person in charge of the section to be audited who will either agree to it or negotiate for a change. An example of an audit schedule is shown in Fig. 5.2.

Before the day of the audit, each auditor is to familiarize himself with the documented procedure(s) involved in his part of the work. He must know who is responsible for what and what quality records to keep. To make the audit run smoothly, he should prepare a checklist showing the following:

- persons to interview;
- questions to ask of each person;
- documents to check;
- quality records to examine;
- in case of a site audit, areas of site and materials to inspect.

The importance of preparatory work cannot be over-emphasized. Failure to do so is recipe for chaos. Without the appropriate checklist in hand, the auditor would just wander around and conduct some casual questioning. At the end of the day, uncoordinated clues have been collected while key information is missing.

### 5.4.3 Process of quality audit

An internal quality audit commences with an opening meeting of the audit team with the person in charge and his immediate subordinates. In the meeting, the lead auditor elucidates the purpose and scope of the audit. He also reviews any nonconformity found in the previous audit, the corrective action taken and its effectiveness.

**Quality Construction Ltd.**
**Building Contractors**

### SCHEDULE OF INTERNAL QUALITY AUDIT

Audit No. : 98/4

Audit date : 23 March 1998

Audit location : Sincere Insurance Building

Auditor(s) : Ms. Amy Beauforte (AB) - lead auditor
Mr. Charles Denson (CD)

Auditee : Mr. Tony Barlowes - project manager

| Time | Activity | Staff to be present |
| --- | --- | --- |
| 0830 - 0900 | Opening meeting | Project manager<br>QA officer<br>General foreman |
| 0900 - 0930 | Tour of site | QA officer |
| 0930 - 1030 | Review of previous audit | QA officer |
| 1030 - 1200 | Document and data control, updating of quality plan (AB)<br>Material handling and storage (CD) | QA officer<br>Site clerk<br>General foreman<br>Storekeeper |
| 1300 - 1430 | Purchasing, subcontractor assessment, equipment maintenance and calibration (AB)<br>Process control, inspection and testing (CD) | QA officer<br>Assistant quantity surveyor<br>General foreman |
| 1430 - 1500 | Project-specific procedures | QA officer<br>General foreman |
| 1500 - 1530 | Nonconforming work, corrective action | QA officer |
| 1530 - 1630 | Preparation for audit summary | |
| 1630 - 1730 | Closing meeting | Project manager<br>QA officer<br>General foreman |

Prepared by  *Amy Beauforte*   Date  9 March 1998
                Auditor

Agreed by   *Tony Barlowes*    Date  16 March 1998
                Auditee

**Fig. 5.2** Typical schedule of internal quality audit.

## 78 Implementing a quality system

After the opening meeting, the auditors go about their work separately, following the checklists they have prepared. By way of interview, observation and inspection of records, the auditors look for evidence of the following:

- defined responsibility and authority;
- awareness of the quality requirements and the role each individual has to play;
- knowledge and understanding of the documented procedures and work instructions;
- degree of adherence to the documented procedures and instructions;
- effectiveness of the quality system.

To assess how closely a documented procedure is followed, an auditor has to talk to the person(s) actually performing the task. (It is far too often that an audit is performed solely in the manager's office.) He should cross-check between people engaged in the same activity, and even request an operator to carry out the activity in his presence. Sometimes the procedure involves different sections of the company, and the audit has to follow the loop to the end.

Nonconformance exists if it is found during the audit that an activity does not comply with planned arrangements. Nonconformities commonly found are:

- unclear responsibility and authority;
- documented procedure not implemented;
- documented procedure not adhered to;
- obsolete documents not removed from points of use;
- marked-up changes made on controlled documents (especially drawings) by an unauthorized person;
- nonconforming work repeatedly occurring;
- cause of nonconforming work not investigated;
- insufficient quality planning of project;
- insufficient training of personnel;
- numerous client complaints.

More often, nonconformities of a less serious nature are noted in the audit. It may be the intention to follow the documented procedure, but due to human error or oversight a certain step has been omitted, e.g. a quality record not signed off or a measuring instrument overdue for calibration, etc. However, when a minor nonconformity happens time and again, it becomes a major nonconformity.

It should be remembered that the purpose of the audit is to verify conformance, not to look for nonconformance. It is a misconception that nonconformities are bound to exist; and much effort is unnecessarily diverted to fault-finding. Even if a nonconformity is noted in the process, it is not the job of the auditor to approve or disapprove the operation as performed. Nor is it his job to resolve the problem discovered. What he has to do is to record his findings faithfully and impartially.

On completion of evaluation, it is usual for the audit team to hold a closing meeting with key members of the audited section. In this meeting, the lead auditor presents a verbal summary of the findings and invites the auditees to respond. This is followed by a brief discussion on the appropriate corrective action. The responsibility for improvement lies with the managing staff of the section audited. Suggestions for corrective action would come from them or be agreed by them. It should be noted that rectifying the actual nonconformity, although necessary, may not by itself prevent recurrence. Any corrective action must get to the source of the problem. If the corrective action cannot be hammered out on the spot, the lead auditor will set a time limit for the auditee to make a proposal.

Before leaving, the audit team usually issues a corrective action request, indicating the nonconformities noted and the corrective action(s) to be taken. This can be done on a standard form as shown in Fig. 5.3. Space is provided on the form for the auditee to signify his agreement. (If a documented procedure is found to be impractical or inefficient, the corrective action request to amend the procedure is directed to the quality manager.) A provisional date is set for a follow-up audit if it is required. The follow-up audit is to verify and put on record that the corrective action has been taken and it is effective. It can vary from a short visit to a complete re-audit and is carried out one month to six months later.

If only minor nonconformities are found, it may be left to the auditee to make improvement in due course. The situation will be reviewed in the next scheduled audit.

After the audit is completed, the lead auditor submits a report to the quality manager with a copy to the auditee. The report normally contains the following:

- audit report number;
- date and place of audit, with name of person in charge;
- names of auditors;

**Quality Construction Ltd.**
**Building Contractors**

### CORRECTIVE ACTION REQUEST

Audit No. :                 *98/4*

Audit date :               *23 March 1998*

Audit location :         *Sincere Insurance Building*

Auditor(s) :               *Ms. Amy Beauforte (AB) - lead auditor*
                              *Mr. Charles Denson (CD)*

Auditee :                  *Mr. Tony Barlowes*

| Nonconformity noted | Corrective action | Target date for action |
|---|---|---|
| 1. *Quality plan not updated after changes were made* | *Project manager to review quality plan at monthly intervals* | *Immediate effect* |
| 2. *An obsolete drawing being used by electrical subcontractor* | *Quality engineer to verify and record return of superseded drawings (See Quality Procedure No.12.)* | *23/4/98* |
| 3. *Outdated cement bags not properly labelled and segregated in store* | *General foreman to inspect storage area once a week and enter the action into the inspection record book* | *Immediate effect* |
| 4. *No supervision of concreting of fifth floor on day of audit* | *General foreman to assign foremen to specific activities on day-to-day basis and to show assignment on notice board in site office* | *Immediate effect* |

Follow-up audit required ?    [   ] Yes      [ ✓ ] No

If yes, provisional date of follow-up audit  ………………………

Prepared by    *Amy Beauforte*      Date    *23 March 1998*
                            *Auditor*

Agreed by      *Tony Barlowes*      Date    *23 March 1998*
                            *Auditee*

**Fig. 5.3** Typical corrective action request.

- scope of audit, showing quality elements checked;
- summary of observations;
- nonconformities noted;
- corrective action(s) requested;
- outcome of follow-up audit, if performed.

Interpretation of a quality audit report is sometimes controversial. For example, a report which discloses a significant number of nonconformities, major and minor, could be taken as a hint that the quality system is ineffective or, on the contrary, as a demonstration that the auditing process is working well. Nevertheless, if the same nonconformities keep on surfacing in consecutive audits, it is without doubt that the quality system is far from being perfect.

To summarize, the audit process is carried out in four distinct but equally important phases: planning, preparation, execution and reporting. A lot of effort is spent on planning and preparation which together occupy approximately 40% of the time. The actual audit exercise may take up 30% of the time, and the reporting work that follows (including follow-up audit if necessary) is done in the remaining 30% of the time.

Based on the outcome of the internal quality audit, the quality manager makes amendments to certain quality procedures if considered appropriate. The reports of different audits are summarized and brought forward at the management review meeting when it is next held.

## 5.5 MANAGEMENT REVIEW

Senior management consisting of directors and section managers should review the quality system at regular intervals, usually once a year. This is to ensure that the quality system continues to be suitable and effective in satisfying the quality policy and objectives of the company. Reports of internal quality audits, and report of external audit if available, are brought up in the review. On occasions it may be found that the quality system has to be modified to cater for changes of the company's business direction, the clients' needs or government regulations. Effects of amendments made to the quality system in the last review are also assessed. Other items to be considered include quality cost and on-time delivery.

The management review meeting is normally chaired by the chief executive of the company with the quality manager acting as secretary.

## 82  Implementing a quality system

Minutes of the meeting are recorded, showing the follow-up actions, the persons responsible and the target dates for implementation.

Management review is an important quality activity. It shows the management's continued commitment to quality. It leads to improvement of the quality system. Furthermore, it keeps the system adaptable to the changing business environment, thereby making it a 'live' system.

At the project level, a similar review is conducted by the project team. The project management review is usually held every four or six months and whenever a major quality deficiency is found. The review covers mainly the results of internal quality audits, changes of client's requirements and/or government regulations, progress of the project, provision of resources and needs for staff training. The project quality plan is updated accordingly and then reissued.

In the project management review meeting, the project manager usually acts as chairperson and the quality officer (or quality engineer) on site acts as secretary. Some subcontractors may be invited to the meeting if they are involved in an incident of nonconforming work. The review meeting is minuted. A copy of the minutes is forwarded to the quality manager who will include it in his report to the corporate management review meeting when it is next held.

## 5.6  SUMMARY

- A quality system normally takes three to six months to implement. During this period, the system documents are amended in the light of the experience acquired.
- A quality system relies on people to make it work. Staff at all levels should be motivated and trained to operate the system.
- Functioning of a quality system is monitored through a series of internal quality audits coupled with periodic management reviews.
- Internal quality audits should be well planned and prepared. The auditors, drawn from the company's own staff, should have no connection with the activities being audited. Normally they require prior training in quality auditing techniques.
- Planning and execution of an internal quality audit follow the flowchart in Fig. 5.4. Appropriate corrective action is taken when necessary. A follow-up audit may be conducted to verify that the corrective action is implemented and that it is effective.

- The quality system is reviewed at regular intervals to ensure that it continues to be suitable and effective. Management review is carried out both at the corporate level and at the project level.

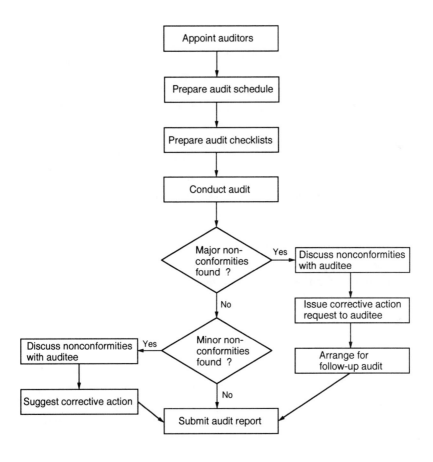

**Fig. 5.4** Internal quality audit.

# 6
# Third party certification

## 6.1 QUALITY SYSTEM AUDITS

A quality system can be audited at three different levels for different purposes. The three levels are: first party or internal audit, second party or customer audit, and third party or certification audit.

As described in Chapter 5, a first party (internal) audit is conducted by a company's own staff to verify whether its quality system is fully implemented and effective. This is to ensure that what you say you would do is actually done and produces good results. Internal audits unveil nonconformities, discover impracticalities and, coupled with periodic management reviews, lead to improvement of the quality system.

A second party audit is carried out by a potential client or his agent on a contractor or supplier. The purpose of the audit is to look for objective evidence that the quality system of the contractor/supplier is effective in ensuring that the product or service satisfies the client's requirements. This type of audit is performed with the consent of the auditee and prior to offer of the contract. After the contract has been entered into, it is not an inalienable right of the client to conduct the audit again unless it is so stipulated in the contract. In fact, it is costly and time-consuming to repeat the exercise. If the client wants assurance that the contractor's/supplier's quality system continues to function well, he may specify as a contract requirement that the quality system be certified by a third party and rely on the surveillance by the third party to ensure that the quality system is working properly all the time.

A third party audit is carried out by an organization with no business connection with the audited company. Essentially the audit is to evaluate whether the company's quality system complies with the stated quality system standard, such as ISO 9001, ISO 9002 or ISO 9003, and whether the system is fully implemented. After the audit, with corrective action taken to rectify any deficiency or nonconformity found, the company is put on a register maintained by the auditing organization and a certificate is issued to the company to this effect. A third party audit is often referred to as third party certification or third party registration, and the auditing organization which performs the audit is called a certification body or a registrar. How widely the certificate is recognized nationally or internationally depends on the recognition enjoyed by the certification body. To achieve credibility of the certificate it issues, the certification body must itself be *accredited* by a national or international organization to be capable of the highly skilful work.

## 6.2 WHY THIRD PARTY CERTIFICATION ?

There are good reasons for seeking third party certification or registration, although it is not mandatory. First of all, it increases the confidence of potential clients in the company's capability of delivering quality products and services. (Anyone is more than willing to entrust one's building work to a quality-endorsed company.) More and more clients are inclined to deal exclusively with companies so certified and make this a pre-qualification for tender. Those companies not falling in line will be ousted from the competitive market in the foreseeable future.

The certificate enjoys national, and maybe international recognition. This enhances the image of the company and may be used as a marketing tool for expanding the company's business to other geographical areas.

Certification provides an opportunity for the company to have an independent evaluation of its quality system. A team of auditors from an organization with no connection with the company will assess the quality system with an objective mind. They are often able to point out some deficiencies in the quality system which have been overlooked.

Although certification is not the end of the quality journey, it is an important milestone in the company's history. This is an achievement that everyone in the organization can be proud of. The morale of the staff is thereby enhanced.

## 6.3 SELECTING A CERTIFICATION BODY

The certification process starts with selection of a certification body or registrar. In most countries, there are governmental and/or private organizations offering services of quality auditing; these organizations establish and maintain registers of companies which have passed their audits. Certification bodies with worldwide branches include Lloyds Register Quality Assurance, Det Norske Veritas, Bureau Veritas Quality International, SGS International Certification Services, TUV Rhineland and others. Some certification bodies which serve mainly local regions are British Standards Institution Quality Assurance, Standards Australia Quality Assurance Services, National Association of Testing Authorities (NATA, Australia), Hong Kong Quality Assurance Agency and CIDB-SISIR (Singapore). Certification (or registration) of a company by an *accredited* certification body represents acceptable proof within the country, and in any country that has granted mutual recognition of such certificates, that the company has a quality system in operation conforming to the stated Standard.

For its certificates to be recognized nationally and internationally, a certification body or registrar has to pass an evaluation by an 'accreditation body' set up by the government. Accreditation ensures that the certification body is competent, impartial and free of conflicting interests. In the United Kingdom, the accreditation is handled by the United Kingdom Accreditation Service (UKAS)[*]. Other EU nations have accreditation bodies similar to UKAS. In the Australasian region, the Joint Accreditation System for Australia and New Zealand (JAS–ANZ) serves the same purpose. The equivalent in the USA is the Joint ANSI/RAB American National Accreditation Program for Registrars of Quality Systems, although this is not a government sponsored scheme.

The national accreditation bodies of different countries have been negotiating with each other to work out recognition agreements on a bilateral or multilateral basis. With mutual recognition, a certificate issued by an accredited certification body in one country is also recognized in another country.

In a country or region, there may be a number of certification bodies in practice. For example, there are at present more than ten organizations offering third party certification in the Australia–New Zealand region. A certificate issued by one body has the same status as a certificate issued by another body, if both certification bodies are accredited. However, not all certification bodies have auditors with experience in your field of activities. It may not be essential for the auditors to possess working experience in the construction industry, but a good knowledge of the

---

[*] formerly the National Accreditation Council for Certifying Bodies (NACCB)

construction operations will enable them to interpret the requirements of the quality standard in the context of the conventional practice. Besides, the time to reach certification is partly dependent on how soon the certification body can start processing your application and how much human resources can be allocated to it. Furthermore, the initial charge and ongoing charge for certification vary from one certification body to another. Therefore, in selecting a certification body, the following aspects need to be considered:

- Credibility   Is the certification body accredited? Are their certificates recognized nationally (and internationally)?
- Experience    Has the certification body audited companies similar to yours?
- Services      Does the certification body offer other services such as training of internal quality auditors?
- Resources     Does the certification body have a sufficiently large pool of qualified auditors to draw upon?
- Availability  How soon can the certification body process your application?
- Charges       How much does the certification body charge for initial evaluation and subsequent surveillance to maintain certification?

## 6.4  SCOPE OF CERTIFICATION

Before applying for certification, a company has to decide on, and to define, the scope of certification that best suits its operational needs. The scope of certification should clearly specify the following:

- standard to which quality system is certified;
- scope of work to be covered;
- divisions or branch offices to be covered.

Normally the quality system of a building construction company is to be certified to either ISO 9001 or ISO 9002 depending on whether the company undertakes design as well as construction. However, the company may elect to qualify to a modified scope by adding to or subtracting from the scope of the Standard.

The company must also indicate the scope of work to which its quality system is applied. The scope of work of a building construction company may generally be described as 'building construction and ancillary works'. It can be more specific as to the type of buildings (e.g. residential or commercial buildings, steel or reinforced concrete buildings) and the construction activities (e.g. general building work, erection of steelworks, piling or building services). As the work of a building construction company is varied and unpredictable, it is advisable to keep the scope as broad as possible. It should be remembered, however, that in broadening the scope of certification, the company is at the same time broadening the scope of evaluation of its quality system. Under certain circumstances, it may be to the advantage of the company to limit the scope of certification to part of its operations. For example, a company undertaking work for the government as well as the private sector can arrange the certification to encompass the government projects only. Another example is a company whose business involves construction of new buildings as well as refurbishing of old buildings; such company can confine the certification to sites of new construction.

Certification normally covers the entire organization and all construction sites. There is a risk, however, that failure of one division or a construction site to comply may delay the award of the certificate or, if occurring after certification, result in suspension of certification.

## 6.5 PROCESS OF CERTIFICATION

The process of certification is illustrated by the flowchart in Fig. 6.1. It comprises the following steps:

- application;
- document review;
- pre-audit visit;
- system audit;
- follow-up audit, if necessary;
- registration and award of certificate;
- surveillance audits.

Following the decision to go for third party certification, the company should approach an appropriate certification body as early as possible.

Although the certification body cannot act in the capacity of a consultant because of possible conflict of interests, it may advise the company on the adequacy of the quality system and the readiness of the company for certification. Most certification bodies offer a preliminary and informal evaluation service at this stage which is, however, optional.

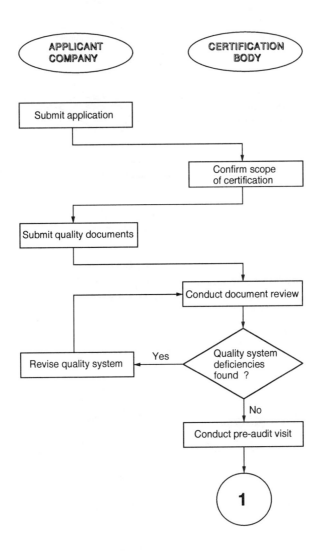

**Fig. 6.1(a)** Process of certification.

## 90  Third party certification

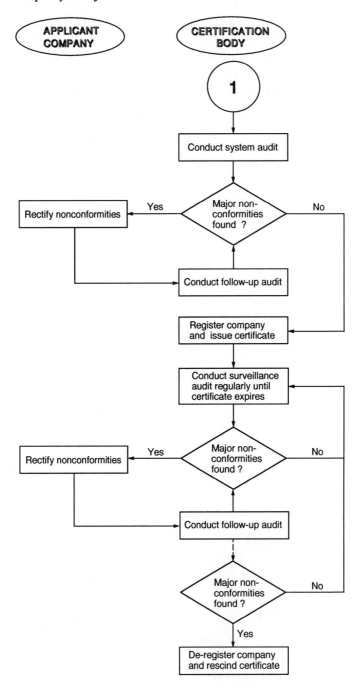

**Fig. 6.1(b)**  Process of certification (contd).

The certification process is initiated with the submission of an application and payment of an initial fee. There is usually a form to fill in on which the applicant is to provide the following information:

- nature of business;
- number of staff;
- locations of work areas / branches;
- scope of certification;
- stage of development and implementation of the quality system.

Most certification bodies or registrars also require the applicant to respond to a questionnaire from which the company's readiness for certification can be gauged. If the company does not appear to be on the right track, the application is temporarily withheld while the company tries to improve its quality system. The company may seek the assistance of a quality management consultant other than the certification body to which application has been made.

On receipt of the application, the certification body places it in the hands of one of its officers. He/she will contact the company's quality manager and confirm or clarify the scope of certification. Usually this officer will later be the lead auditor of the audit team. The audit team formed will include at least one member with some knowledge and experience in the business of the audited company, in this case the construction industry.

Once the application is accepted, the certification body will request the company to submit the documentation of the quality system for document review. The essential documents are the quality manual and the quality procedures. They will be thoroughly checked to ensure that every relevant element of the stated Standard has been satisfactorily covered. It is not uncommon to find that some requirements are incorrectly interpreted or inadequately addressed. The company is then informed of the deficiencies in the quality system that need to be rectified before a system audit is conducted. In particular, if the quality system is so inadequate that it requires a major overhaul, the company would be advised to develop it further before resubmission. The revised documentation must be judged as satisfactory before the system audit can proceed.

In conjunction with the document review, or immediately after, the officer-in-charge normally makes an informal visit to the company. The purpose of the pre-audit visit is two-fold: firstly it enables the auditor to appraise the physical environment of the company (including construction sites) where the audit is to be conducted; secondly it gives the auditor a glimpse of the quality system in action. The broad overview so acquired will assist the audit team to assess the length of time required for the system audit and to work out the audit schedule. During the pre-audit visit, the auditor might identify an obvious quality deficiency and alert the company's management to the need of corrective action before the formal audit.

Some dates are set for the system audit on the company's premises and construction sites. The audit normally takes two to four days depending on the size of the organization and the complexity of the quality system. It is carried out by an audit team of two or three headed by a lead auditor. The company receives an audit schedule in advance. The schedule provides details of the time of visit to the various work areas and the persons expected to be present.

The system audit is similar to the internal audit described in section 5.4, except that this time the entire quality system is evaluated throughout the entire organization. The audit begins with an opening meeting, followed by the auditing work (usually with a review meeting of the auditors at the end of each day), and finishes with a closing meeting. During the visit to the various work areas, the audit team would look for objective evidence that the quality system is effectively implemented. Following the checklists prepared beforehand, individual auditors would observe operations, interview operators and inspect quality records, so as to verify whether the quality activities and related results comply with the documented procedures and quality plans. They often check the drawings and other documents on site to make sure that no obsolete ones are in use or on display. They even ask for the work instruction sheets which, according to the quality plan, are issued to those performing special tasks, and then compare what is written thereon with their observations. In simple words, they want to find out whether 'you do what you say you do'.

Although the quality system would have been put in practice for some time when the audit is conducted, do not expect the auditors to find everything functioning according to plan. The auditors would probably discover that some documented procedures are not strictly adhered to or some quality records are missing. They might also notice that some quality related activities have no documented procedures or work instructions. (A major fault commonly found is poor document control especially on site.) All nonconformities observed are then sorted, classified and made known to the company management. The audit report brings out the positive as well as negative points of the quality system but does not specify any corrective action. (It is up to the company to seek specific advice from an independent consultant, should it be considered necessary.) The subsequent action is dependent on how serious the nonconformance is. With major departures from the documented procedures, the company is usually given one or two months to respond, after which time a follow-up audit is conducted by one of the auditors to verify that the corrective action has been taken and it is effective. On the other hand, minor and isolated nonconformities will not hold up certification. The company is required to take appropriate action to prevent recurrence of nonconformance and the situation will be reviewed during the surveillance audit.

Depending on the results of the system audit, and the follow-up audit if performed, there are three possible outcomes:

- Certification:
  The company is granted unconditional certification.

- Conditional or provisional certification:
  The company is certified but is also required to attend to some minor nonconformities within a time frame set by the certification body.

- Refusal:
  The company is denied certification and the grounds for refusal are given.

If certification is granted, the company is entered into a register of certified companies kept by the certification body and a certificate is issued to the company. The certificate shows the name of the company (and the division if applicable), the Standard certified to, the scope of certification and the date of expiry. It bears a unique number and a 'quality mark' of the certification body. The company may incorporate the mark on its letter paper and other items of stationery *but not on its products*. The company may also use the mark in advertising.

For the company, attaining certification is hard work; keeping it can be even harder. After the system audit is over, some staff tend to fall back onto their bad working habits. Management should continue to show its commitment to quality and to lead the organization on the long journey of quality improvement.

## 6.6 PREPARING FOR SYSTEM AUDIT

To the company, the most important step in the entire process of certification is the system audit conducted on its premises and construction sites. Careful preparation for the occasion will pay large dividends.

The system audit is probably the first time that the staff of the company are exposed to scrutiny and questioning by an external team. It is natural that they feel somewhat nervous in front of the auditor. Some effort should be made beforehand to put them at ease.

As anyone in the company could be caught up in the audit, all members of staff irrespective of rank should know what to do when questioned by the auditor. The quality manager may brief the department heads and managers (including project managers on site) who will pass on the message to their subordinates. It is especially important to make clear that the audit is on the quality system, not the staff. One should respond honestly and cooperatively to the auditor's queries but does not need to offer information beyond that asked for. A common mistake is to answer, often incorrectly, something outside one's area of responsibility.

During the system audit, the audit team will be guided around the premises. One or two members of staff should be pre-selected to serve as guides. Ideally they are persons who have served as internal quality auditors and performed well in the task. They are there to facilitate the work of the auditors but not to participate in the audit process at all. They should neither ask questions nor answer questions related to the audit.

An auditor invariably asks for some quality records for inspection. Each section of the company, and each construction site to be visited, should get all quality records handy. It would give the auditor a bad impression if you start searching all over the place for a record when it is asked for. It is also unwise to refuse access to certain records on the ground of commercial secrecy, because by doing so you appear to have something to hide. Likewise, declaring certain areas off-limits to the auditors will create suspicion. In fact, all registered (certified) auditors are bound by a code of conduct to maintain confidentiality of what they observe in performing their task.

While the quality system may be operating well in general, there are areas in which auditors commonly find faults. Special attention paid to these areas could increase the chance of passing the audit in the first round. Experience of auditors indicates that the following five areas are the most vulnerable.

| Quality system element | Examples of nonconformance |
|---|---|
| Document control | superseded documents lying around; changes made to drawings unauthorized. |
| Equipment calibration | calibration of equipment outdated; out-of-order equipment not out of service. |
| Quality planning | project quality plan not prepared; project quality plan not updated. |
| Corrective action | nonconforming work rectified but no corrective action taken; effectiveness of corrective action not verified. |
| Training | staff not properly trained before taking up new duty; on-the-job training not recorded. |

Other than the preparation as described above, business should go on as usual on the day of the audit. No extra effort needs to be spent on anything which is not usually done. The auditors are looking for evidence of *continuous* conformance, not a show put up on the day of audit.

## 6.7 SURVEILLANCE AUDIT

Issuance of the certificate does not signal the end of the certification process. Instead, it is the beginning of a long-term relationship between the two parties. On the one hand, the certification body continues to provide testimony to the company's quality activities; on the other hand, the company agrees to be subject to periodic surveillance audit and to pay an annual fee for the service.

The certificate is normally valid for three years although some certification bodies issue certificates which expire after two years. During this period, the certification body will conduct periodic surveillance audit to ensure that the company's quality system continues to function well and any changes made are in order. The surveillance audit is usually carried out every four or six months. It is less comprehensive than the audit for initial certification, covering selected elements of the quality system on a rotational basis. However, within the period of validity of the certificate, each quality system element will be audited at least once. If, at any time, the quality system is extensively amended, the certification body must be informed immediately. The revised documentation will be reviewed; an additional surveillance audit may also be conducted.

Prior notice of the surveillance audit is normally given and the date is mutually agreed, although it can also be an unannounced visit. In the surveillance audit, attention is focused on the following:

- areas where nonconformities were noted in the last audit;
- effects of corrective actions since the last audit;
- changes in the quality system since the last audit.

The surveillance audit is conducted in more or less the same way as the certification audit. A major nonconformity noted in a surveillance audit calls for urgent corrective action, failing which the certification may be suspended or withdrawn.* A minor nonconformity, though not a pressing issue, should be corrected as soon as possible.

Just before the certificate expires, another system audit is arranged and conducted. Once again, the comprehensive process is gone through. Should the quality system be found operating effectively, certification is renewed for a further period.

## 6.8 TIME AND COST OF CERTIFICATION

The time required to achieve certification after lodging an application varies with the following factors:

- size of organization;
- scope of certification;
- complexity of the quality system;
- maturity of the quality system in terms of development and implementation.

Document review by the certification body normally takes two or three months. If some deficiencies are found, which is not unusual, revision of the quality documents and re-examination by the certification body can take up another three months. In the meantime, a pre-audit visit may be arranged. After the desk-top audit, the on-site audit could be expected

---

* Certification may also be suspended or withdrawn as a result of misuse of the quality mark provided by the certification body.

within the next three months. It is unlikely that the company will pass the comprehensive system audit without a hitch. Corrective action and follow-up audit may then cause a delay of yet another three months. Altogether the time taken to achieve certification is seldom less than six months even when everything runs smoothly, and is more likely to be close to twelve months. Some companies have taken as long as two years to get their certificates. Do not rush for certification when the company is not ready, as it will result in unnecessary delays and costs, and definitely frustration. On the other hand, development of the quality system does not stop with certification: it is a continuing process through which the system is gradually improved.

The cost of certification in money terms depends on the amount of work to be done by the auditors and the unit charges set by the individual certification body. Most certification bodies charge separately for the following services:

- application fee                                  lump sum
- document review                           based on auditor-days
- pre-audit visit                                based on auditor-days
- system audit                                 based on auditor-days
- follow-up audit                            based on auditor-days
- annual registration, including
  surveillance audits                        lump sum
- expenses, such as travel and
  accommodation of auditors         at cost

At the current rates, a rough estimate for a small to medium-sized company would be US$10 000–15 000 for initial certification and US$3000–5000 per year afterwards. This is minimal compared to the internal costs, in terms of staff effort and time, and the consultant's fees put into the development and implementation of the quality system. However, the benefits of being a quality endorsed company will in the course of time pay back the total expenditure many times over.

## 6.9  SUMMARY

- Quality auditing may be carried out by the company's own staff, by a client or by a third party.

## 98  Third party certification

- A third party quality audit is carried out by an independent body that is usually accredited by a national council. After a successful audit, the company is registered and a certificate is awarded.
- Certification or registration by a third party provides formal recognition of the status of the quality system. It enhances the image of the company and the client's confidence in the company. Although it is not mandatory, there is business necessity to do so.
- For a building construction company, its quality system is certified to comply with either ISO 9001 (with design) or ISO 9002. The scope of certification may encompass all or part of the company's business activities, and may cover the entire organization or the divisions / branches specified.
- The certification process comprises a desk-top audit of the documentation of the quality system and an on-premise audit of the quality system in action. Certification is generally valid for three years, and is maintained through periodic surveillance audit.
- The certification process normally takes six months to one year. The cost of certification varies from case to case, but the benefits far outweigh the expenses incurred.

# 7

# Facts and fallacies

## 7.1 PERCEIVED OUTCOMES OF QUALITY SYSTEM

Implementing an ISO9000-compliant quality system in an organization has long-term effects. A SWOT analysis[*] reveals the possible outcomes as follows.

*Strengths*:

- less rework or repair;
- stronger client focus;
- higher efficiency in operation;
- improved internal communication;
- improved external communication;
- systematic record keeping.

*Weaknesses*:

- more paperwork;
- increased bureaucracy;
- higher overhead cost.

*Opportunities*:

- more business locally;
- more business inter-state and overseas;
- returned business from satisfied clients.

---

[*] SWOT stands for strengths, weaknesses, opportunities and threats.

*Threats*:

- staff discontent;
- less flexibility in operation;
- lower productivity in the initial period.

Now that you have established a quality system conforming to ISO 9000, you would naturally expect to reap the perceived benefits. However, the benefits will not crop up overnight. The quality system needs time to evolve and take root in the organization, let alone to bear fruit. Substantial training efforts are required to make the workers become quality conscious. Experience gathered from the manufacturing industry indicates that it takes at least five years for a quality system to become fully operational and effective. Such experience with the construction industry has yet to be collected, but the time would not be shorter. Quality assurance is a concept that rewards patience. Management should look at it as a medium to long term strategy of the company.

A quality system has to be amended and improved continuously, via internal quality audits and management review. The first version would not work to your satisfaction, despite of the fact that it was developed from current practices. The second version is more adaptable and user-friendly. When it comes to the third version, you will see the results that you have hoped for.

## 7.2   CRITICISMS OF ISO 9000

The benefits or otherwise of implementing an ISO9000-compliant quality system in a construction company have been the subject of contention for some time. Some contractors experience a change for the better. Others remain status quo. Yet others degenerate into a state of chaos: complaints abound, both from management and the workforce. Indeed the construction industry has a jaundiced view of quality assurance along the line of ISO 9000. Let us examine the major criticisms of ISO 9000 in practice.

### 7.2.1   Quality system as business necessity

The quality movement in the construction industry has been mostly government driven. Many a contracting company has implemented a quality system in order to qualify for tender rather than arising from any real concern for quality. The certificate is virtually a licence to work. Management considers quality assurance as a window-dressing exercise that creates extra but unrewarded work, cutting the profit margin. To maximize profit, of course, is fundamental to a business. But systematic planning and control leads to high efficiency and productivity. Hence, high quality can be attained at a lower cost.

When a company implements a quality system as a business necessity, it has got the priorities wrong. The company should put quality in the first place and use the quality system as a means of achieving it. The quality (management) system is essentially a management tool. It can be used to streamline the daily operations. With an effective quality system in place, it is a matter of time that the organization is transformed into an efficient and productive workforce, turning out high quality product. If the system does not bring about these outcomes, it is not suitable for the organization.

### 7.2.2 Resistance to change

From the point of view of the workers, it is pointless to document procedures for activities that they have been performing every day. Well, if they can perform the tasks 'right the first time every time', there is no need to write down the procedures. In fact, ISO 9000 states that documented procedures are only necessary where the absence of such procedures could adversely affect quality. It also states that the range and detail of the procedures are dependent upon the complexity of the work, the methods used, and the skill and training needed by personnel involved in carrying out the activity. Simple documentation is the rule.

Implementing a quality system does not alter the current practices in substance, nor does it increase the amount of inspection and testing. What it does is to arrange these activities in a planned and systematic manner. People-on-the-job need not worry about the change in mode of operation which is apparent only. In essence, everything stays put.

The attitude of people towards quality has an important bearing on the functioning of a quality system. The irresponsible performance of a few individuals could jeopardize the whole system. On a construction site where people with different background, skill and education work together, it is difficult, but not impossible, to make them all abandon their long-held habits and follow the established procedures. Most workers are hardly aware of quality assurance and its requirements. To unlock their resistance to change, training is the key. The concept of quality should be cultivated throughout the organization. For this purpose, extra resources (including staff time) must be allocated. The quality drive will succeed if commitment is firm and resources are provided, but will fail miserably when these conditions are not met.

Site staff, and office staff to a certain extent, often complain about the extra workload resulting from the use of various forms and checklists. They claim that time is better spent on performing the nitty-gritty. When forced to do so, they just pay lip service to the paperwork. A tick in the box does not necessarily mean that verification has been carried out. The worst scenerio is that they fill in the forms and checklists *restrospectively*, maybe a few days before a quality audit (either internal or external) is scheduled to take place. This completely violates the purpose for which the paperwork is intended, and thus jeopardizes the effectiveness of the quality system.

Again, training is needed to rectify the malpractice. When people-on-the-job get used to the new routine, they will accept the forms and checklists as a reminder of what has to be done and, when completed, as a proud record of their achievement.

### 7.2.3 Nonconformance – a crime or a lesson?

Reporting and recording of nonconformance is central to the functioning of a quality system. The idea is to avoid making the same mistakes again in the current project or in a future project. Unfortunately, people are accustomed to hiding their own faults and covering up the defective work. Even their supervisor is often relunctant to go by the book. Instead of condemning the work and reporting the case, he or she would rather go on with it, on the pretext of time constraint.

It should be remembered that to err is human. Everyone makes mistakes sometimes, including deviation from a documented procedure. An unintentional fault should not be treated as a crime, unless it is repeatedly committed, and the offender should not be penalized. Instead, it should be dealt with openly as a lesson so that others can learn from it. Adequate resources (including time) should be provided for a job, and everyone involved would have no excuse for cutting corners or dispositioning a nonconformity in an improper or unauthorized way. This is a cost that management must be prepared to bear to practise quality assurance. It will be paid back many times over when the project proceeds without interruption.

The client, or the architect / engineer acting as his representative, would probably be suspicious of the contractor's capability in producing quality work when he sees a long list of nonconformities. In fact, he should begin to worry if he is presented with a clean sheet of paper. A contractor who has the courage to disclose its own errors is unlikely to err again. On the contrary, a contractor who keeps his wrong-doings as a secret is virtually hiding a time bomb on site – liable to explode anytime in the face of the client!

It is to the advantage of both parties to record noncomformance. The client should encourage it. The contractor should do it.

### 7.2.4 Third party auditing

Quality auditing by a third party has been a contentious issue ever since its introduction. It is argued, not without reason, that an audit done by someone external to the company is either superficial or unrealistic. Auditors from a certification body are generally unfamiliar with the construction industry and its conventional practice. As a consequence, they tend to be pedantic and inflexible in their assessment, paying too much attention to the literal meaning of the relevant standard. For the same reason, different auditors may interpret the requirements differently, even when they come from the same certification body. It is general practice now to include in the audit team one person knowledgeable in the industry involved in a particular audit, as recommended by ISO. A company may check on this before engaging a certification body.

The criteria for assessing a quality system are: (a) Does the system comply with the standard to be certified to? and (b) Is the system effective? In going about their work, auditors should use an open mind and a pair of discerning eyes. Experienced and informed auditors would consider the relevance of each requirement of the standard. They would insist on those requirements that bear directly on the company's operations, but would not bother about minor deviations from the less relevant requirements. This is particularly important for a small business. While a large organization requires a huge volume of written procedures to clarify the management structure, authorities, responsibilities, channels of communication and standard methods of operation, a sole propriety or partnership will find such procedures cumbersome and unnecessary. In judging whether the

## 104  Facts and fallacies

quality system is effective, these auditors would be guided by evidence of quality improvement, such as reduction of rework or repair. They would pay more attention to recurrence of nonconformance than to first offence.

Another common criticism of external auditing is the varying degree of persistence of the auditors. Some auditors are strict, demanding immediate corrective action for minor nonconformities. Others are lenient, letting loose the audited company even on major issues. The inconsistent standard of auditing deals a heavy blow on the credibility of the third party certification scheme. Given that ISO 9000 is meant to minimize variability in quality, it is an irony that such inconsistency arises from use of it. To ease the problem, ISO has issued guidelines for auditing, qualification criteria for auditors and management of audit programmes (ISO, 1991). In some countries, an accreditation scheme of quality auditors is operated by a government or semi-government body. Through this, the standard of auditing will gradually be improved.

### 7.2.5 Is the clerk-of-works dispensable?

While some contractors are resentful of the adoption of ISO 9000 in their industry, some clients are unconvinced of the benefits that ISO 9000 is supposed to bring. They find to their disappointment that certain contractors who are quality certified still cannot produce defect-free works. They are doubtful where quality is in quality assurance.

After pre-qualifying contractors on the basis of third party certification, some government departments no longer employ a clerk-of-works on site, believing that the contractor's certified quality system would provide them with enough protection. It ends up with nobody from the client's side checking the construction work. They check the forms instead. The quality of the end product is at the mercy of the contractor.

Should the clerk-of-works be retained on site even when the construction is done by a quality-certified contractor? The answer to this question is probably positive, but the role of the clerk-of-works needs to be redefined. He is to monitor rather than to supervise the contractor's and the subcontractor's work. The part he plays in quality control blends with the contractor's part by way of hold points and witness points of the various inspection and test plans. His presence at these stages of work ensures that the contractor carries out the verification as planned. His signature on the ITP serves as approval or acceptance of the work up to that point. (If there is no clerk-of-works on site, a representative from the architect's or engineer's office has to make frequent trips to the site anyway.) The only difference from the traditional practice is that the inspection work is jointly planned, organized and scheduled. In this way, a sense of partnering is promoted, resulting in cooperation instead of confrontation.

The job of the clerk-of-works usually includes certifying periodic payment which is linked to acceptance of the works completed. Unlike manufacturing where acceptance is conducted in-house by the supplier's staff, the authority to accept in a construction project is vested in the client's representative according to standard forms of contract. Since the supplier cannot accept his own product on behalf of the purchaser, there is a need to provide for purchaser acceptance (Low, 1998).

## 7.3 ISO 9000 IN SMALL BUSINESSES

There has been widespread criticism that ISO 9000 is not suitable for small businesses. The costs of attaining and maintaining certification are too high. The system requirements are inflexible. The paperwork involved is tedious and mostly irrelevant. It is argued that informal measures of quality management are more applicable and cost-effective. Some people even advocate that small businesses should not be required to go for ISO 9000 certification, if they can provide evidence of the consistent quality of their products or services.

As many trades are involved in construction, subcontracting is the rule. A small company often acts as subcontractor to a large company besides serving its direct clients. In a sizable project where quality assurance is specified, the main contractor has to embrace the site activities of the subcontractors in his quality system, or to demand them to operate quality systems that are in line with his system. Under certain circumstances, a subcontracting company may even have to be ISO 9000-certified in order to qualify for the work.

A small business is different from a large enterprise in several ways. The company may range from a sole proprietor, a partnership to an organization employing not more than, say, ten people. It is normally run by one manager who is the proprietor or closely in touch with the proprietor. The manager is knowledgeable and competent in the trade, and is often the only supervisor of the activities. The staff are each responsible for a variety of jobs, but decision-making is confined to one or two persons. As a result, the organization structure is simple and the delegation of authority limited. Authority and responsibility can be simply defined by a simple document showing who does what. Formal job descriptions and organization chart are unnecessary.

In a small organization such as this, the manager usually doubles as the quality manager. Through personal supervision, he exercises process control, reviews and dispositions nonconforming product, and even initiates corrective action when required. Problems related to quality are resolved as they arise and corrective action taken right away. His diary notes serve as records of quality activities. The quality problems are then brought up in the regular meeting of key members of staff, and this constitutes the management review. A simple procedure describing such arrangements is all that is needed to satisfy a number of quality system requirements specified in the standard.

With only a few people in the organization, the route of communication is short. Document control is relatively easy (but equally important). Everyone would have access to a central file of documents, including drawings issued by the architect / engineer or the main contractor. Whenever a document is amended or a revised drawing is received, the superseded document / drawing in the central file is replaced. It is not necessary to establish a master list and a distribution list. Simple as it is, the process of document control should be covered by a documented procedure, indicating the person responsible, the proper handling of the central file and the change control.

The volume of business of a small company is limited and its scope is confined to a small range. The activities of the entire company can be covered by a few documented procedures. These procedures are probably written by people who have been on the job day after day. They have a sense of ownership of the procedures, and are prepared to abide by them every time the job is done, thus assuring consistent quality of the output.

Due to the small size of the workforce, individuals have to verify their own work to avoid duplication of effort. It is not feasible to insist on double check by another person. However, if the architect / engineer or the main contractor feels that independent verification is essential at some critical stages, this can be achieved by establishing hold points and witness points at appropriate places of the inspection and test plan or a similar schedule.

A technical difficulty often arises from the requirement of internal quality auditing. Internal quality auditors should be independent of the activity being audited. In a company of small size, there are only a few people around and everyone is involved in a number of jobs. This makes independent auditing (in the strict sense) impractical. A possible way out is for the entire organization, staff and workers alike, to sit together to discuss about their work and review the quality records. This can be arranged regularly and the results recorded. An alternative approach is to cooperate with another company such that each conducts quality audits for the other (SA/SNZ, 1996).

To sum up, the principles behind ISO 9000 are equally applicable to large and small businesses. As the range of activities of a small business is small, certain requirements of the standard are not applicable. If a requirement can be demonstrated to be irrelevant, a simple statement to this effect in the quality manual would be sufficient to satisfy the standard. With those requirements that are related to the business, there are always simple and practical ways to provide the necessary control. In assessing the adequacy of a company's quality system, auditors from a certification body should make use of the flexibility built into the standard. Helpful hints have been suggested for the business concerned and the assessors (IQA, 1995).

## 7.4 THE WAY AHEAD

Quality should be in everybody's mind. Whether you are a manufacturer of goods or a provider of services, your product has to satisfy the customer. Prosperity of your business relies heavily on customer satisfaction, and following ISO 9000 in managing quality will help you achieve it.

The ISO 9000 family of standards provides the foundation for quality management. It is simply a collection of principles of sound management that have proven their value over many years. The quality system it portrays is based on international consensus and is capable of being tailored to fit any enterprise anywhere in the world.

ISO 9000 is evolving. It started in 1987 with the first issue and was revised in 1994. The second edition has been under review for some time, and by now (September 1998) a draft of the new edition is being circulated worldwide for discussion. Publication of the revised standards is planned for the second half of the year 2000, although any delay would postpone it to the next millennium.

In the forthcoming revision, the current family of some 20 standards will be re-structured and consolidated. Besides, the new ISO 9000 standards will be made more compatible with the ISO 14000 standards for environmental management systems. This is to ensure that common elements of the two families of standards can be readily implemented together without duplication or contradiction. The outcome will be a set of four new standards as follows:

ISO 9000: 2000   Quality management systems – Concepts and vocabulary
ISO 9001: 2000   Quality management systems – Requirements
ISO 9004: 2000   Quality management systems – Guidelines
ISO 10011: 2000  Guidelines for auditing quality systems

The provisions of ISO 9002: 1994 and ISO 9003: 1994 have been incorporated into the new ISO 9001: 2000 (ISO, 1998). Those organizations with quality systems certified to the current edition of standard need not overhaul their quality systems or substantially re-write the documentation. They can transition to ISO 9001: 2000 through limiting the scope of application, or by tailoring the requirements.

In the evolution of ISO 9000, there has been a general shift towards total quality management (TQM) which is a management philosophy emphasizing on quality, teamwork and decisions based on data. TQM advocates an organization-wide effort in continual quality improvement. To attain high quality in construction, all parties involved, ranging from the client, the architect, the engineer to the contractor, the subcontractor and the material supplier, must work together as a team. The practice of TQM promotes good relationships both within the individual organizations and between organizations.

The pursuit of quality is a gradual progression from quality control, through quality assurance to total quality management, the realm of each expanding beyond the previous as illustrated in Fig. 7.1. Implementing ISO 9000 is an important part of the quality journey. Whoever is still at the crossroads should not hesitate any longer. Summon your staff, take the lead, and march towards the goal!

## 7.5 SUMMARY

- The benefits of quality assurance may take years to materialize. In the mean time, the quality system keeps on evolving.

- Most criticisms of ISO 9000 applied to construction are due to people's inertia and misconceptions. Training is essential for a quality system to become effective.
- ISO 9000 is applicable to both large and small businesses. For the latter, the system requirements may be tailored to suit. A flexible approach by the company and its assessors would make certification achievable and worthwhile.
- ISO 9000 is an evolving family of standards. Revision is made every few years. The next issue is scheduled for the second half of year 2000. Despite the proposed changes in structure and arrangement of contents, the fundamental requirements remain practically unchanged. It will not be necessary to re-write the quality system documentation to comply with the revised standards.
- ISO 9000 is moving from quality assurance towards total quality management. The practice of TQM should be the goal of every company that values the satisfaction of its customers.

**Fig. 7.1** Expanding realm of quality management.

# PART TWO

# DOCUMENTING A QUALITY SYSTEM

# 8

# Writing quality system documents

## 8.1 DOCUMENT LAYOUT AND FORMAT

There are no hard and fast rules governing the layout of quality system documents. Whatever arrangement is adopted must present the contents clearly and unambiguously. Besides, the writing should be simple and straight-forward.

A page identification system is necessary to positively identify each page of a document. It normally consists of the following elements:

- company name;
- document number and/or title;
- page number of total number of pages (e.g. *Page 2 of 30*);
- issue number;
- revision number, if appropriate;
- date of issue / revision.

An amendment sheet is normally included in the document for recording the details of amendments, such as the revision number, pages replaced, contents amended, date of effect and signatory evidence of updating. This page is usually placed immediately after the title page.

Quality system documents are live documents and the format should make it easy to update. The documents can be in hard copy form or electronic form. With the hard copy, loose-leaf binding is most suitable as it allows replacement of some pages without disrupting the whole volume.

Putting the quality system documents in electronic media has distinct advantages, despite its initial outlay. The documents are stored as read-only files and made available to users in the office and on site via a computer network. Currency of a document is assured, as the obsolete version is instantaneously removed. Updating of a document is simple but under strict control. Only the authorized person with the security code has access to a file to make changes. Furthermore, the forms attached to the procedures can be handled with a database. Data and information are keyed into the pre-programmed format and are readily retrieved when required. If properly coded, entries from different projects may be compiled, compared or analysed, thereby exposing hidden trends that are otherwise difficult to detect.

## 8.2 WRITING THE QUALITY MANUAL

The quality manual lays down what the company commits itself to do to achieve quality. It should demonstrate that the quality system satisfies the standard with which compliance is intended. The essential elements of the quality system are:

- quality policy statement;
- description of responsibilities, authorities and interrelationships of personnel whose work affects quality;
- list of quality system procedures and work instructions;
- a statement indicating the control, review and updating of the manual.

The quality manual may be conveniently arranged as follows:

- title page;
- record of revisions;
- list of contents;
- foreword (may include the statement for reviewing, updating and controlling the manual);
- quality policy statement;
- organization structure (organization chart plus position descriptions);
- quality functions with reference to quality procedures, or simply a list of quality procedures;
- project quality management.

### 8.2.1 Quality policy statement

The top management's commitment to quality is expressed explicitly in a well thought-out quality policy statement. A bold committal statement shows the management's determination in the pursuit of quality. However, the management must always be ready and willing to stand firmly behind the policy it publicizes. This is the way to win the confidence of the clients as well as the company's own staff.

The quality policy statement is a proclamation signed by the chief executive of the company. It should be straightforward and positive, bearing directly on the company's quality objectives. It may be just a single paragraph like this:

> It is the policy of Excellent Contractors Limited to fulfil the needs and expectations of the clients in the construction contracts it undertakes. To ensure consistent quality of work, the Company establishes and maintains a quality system meeting the requirements of ISO 9002:1994. Adherence to the documented procedures at all times is mandatory for all staff in the Company.
>
> Signed    ...............................
>         General Manager
>
> Date    ....................

The quality policy statement may be a more elaborate document incorporating a brief company profile, thus producing some advertising effect. An example reads as follows:

> High Quality Construction Incorporation specializes in a wide spectrum of building work, covering prestigious office buildings, luxurious residential blocks, modern shopping complexes, multistory car parks and low cost housing. The Company also undertakes design in a design-and-construct contract. It maintains a motivated workforce with appropriate skills and experience to satisfy the needs and expectations of the clients.
>
> Management of the Company firmly believes that client satisfaction is the only safeguard of continuation of the Company's business. The prime objective of the Company's operations is to execute the contract in a manner that satisfies the specified and regulatory requirements. To achieve this objective, a quality system conforming to ISO 9001:1994 is established and maintained in the organization.
>
> The Company's quality system is fully documented in the quality manual and quality procedures. All personnel of the Company shall follow the documented procedures in their activities. Quality deficiencies shall be honestly reported to the appropriate authority who will allocate priority to the corrective and preventive action necessary.

As chief executive of the Company, I urge every member of staff irrespective of position and rank to cooperate with the quality manager in implementing this quality policy.

Signed    ..............................
          Managing Director

Date      ..................

Another example of quality policy statement is shown in the sample quality manual in section 9.2 (page 130).

### 8.2.2 Organization structure

For the staff to act effectively as an integral body, it is essential to define 'who does what' in the organization. Therefore, the quality manual should clearly identify responsibilities, authorities and interrelations. This is conveniently achieved by means of an organization chart (or a number of charts) and position descriptions.

An organization chart defines key management functions and the lines of reporting, from the chief executive of the company down to the person-on-the-job. It encompasses all departments / divisions and all locations where the quality system operates. For a building construction company, the organization chart should preferably indicate the interrelation between office staff and site staff.

A position description in the quality manual specifies the contributions that the incumbent of the position is expected to make in the operations of the company. Typically it is arranged as follows:

Position:
Department / Section:
Reporting to:
Responsibility and authority:
(List here the duties of the person occupying the position. Indicate in each case whether he/she has full executive power or must act in collaboration with another.)

It is not advisable to attach names to the positions on the organization chart(s) in the quality manual. People tend to move around, and the inclusion of names will lead to frequent changes of the document.

### 8.2.3 Quality functions

This section of the quality manual sets out what the company aims to achieve and how. These *self-imposed* rules generally follow the 'quality system requirements' stipulated by the International Standard to which the present quality system is to conform. However, not all twenty quality system elements are applicable to the organization. For example, a construction company normally does not use statistical techniques in verifying the quality of its output and the corresponding requirement is therefore irrelevant. Servicing is another system element that may be irrelevant unless the company provides regular maintenance of the completed works.

The quality functions should preferably be itemized and arranged in the same sequence as the corresponding clauses in the Standard. This facilitates cross-referencing with the Standard, especially in audits by an external body. If any system element is not applicable, it should be indicated as such in the quality manual.

Description pertaining to each quality system element does not need to be lengthy; a paragraph or two will generally be adequate. For details, reference is made to the respective quality procedures bound in another volume. An outline of the structure of the documentation used in the quality system should be included in the quality manual, together with a list of quality procedures.

### 8.2.4 Project quality management

The quality manual should also indicate the measures for project-specific quality management. As building activities vary substantially from project to project, a quality plan is usually prepared to cover the particular needs of a project. A quality plan is a document defining how the quality requirements will be met in a particular project. It may simply be a collection of relevant quality procedures of the quality system, but more often it includes some quality procedures and work instructions written specifically for the project. For example, in a joint-venture project, a quality procedure has to be established for inter-communication among the partner companies. In another project involving diaphragm walling, a work instruction is required to describe the special techniques involved.

## 8.3 WRITING THE QUALITY PROCEDURES

A procedure is an established method for a specific process, describing step-by-step the activities that constitute the process and indicating the persons responsible for these activities. Procedures are in two categories: system procedures and technical procedures. System procedures are used to rationalize the administrative processes such as document control, contract review and internal quality audit. (These procedures are the basis of the quality system and are often referred to as quality procedures.) Through the system procedures, the authorities and responsibilities for particular processes are assigned to individuals (or groups) and the interfaces between them are specified. Technical procedures are prepared, where necessary, to standardize the construction and installation processes and are more appropriately called method statements. Compared to system procedures, technical procedures are more detailed in describing the work involved and less so in addressing the interfaces. Examples of technical procedures are for site formation, concrete work and installation of electrical services. Technical procedures may also be required for verification processes such as weld inspection and pile testing.

Quality (or system) procedures are written for most, if not all, of the quality system elements, providing details on how the requirements are met. For some elements, more than one procedure is needed to cover the quality functions involved. An example is 'contract review'. Different procedures of review are necessary to deal with tendering, contract signing and contract variation. Another example is 'inspection and testing' that is performed at various stages of construction. Procedures are established for receiving inspection of materials, in-process inspection of construction and final inspection of finished works.

A quality procedure should be based on the company's current practice that has been proven practical and effective in assuring quality. The objective is to document such practice, establishing it as a standard method to be followed whenever the process is carried out. It virtually boils down to 'write what you do; do what you write'.

Despite the proven nature of the current practice, it may fall short of the requirements stated in the Standard. In writing a procedure, therefore, you should scrutinize the steps thoroughly to see if they will bring about the expected outcome, and make some minor changes if necessary.

Although procedures are meant to be for internal use, their existence in the organization may have legal implications as pointed out in a CIRIA report (Barber, 1992): 'Failure to follow procedures is a common basis for proving negligence. If a defendant has established such procedures, whether as part of a formal quality system or otherwise, failure to follow them leads automatically to an inference of negligence. Liability may result, irrespective of whether the quality system has been contractually invoked. Procedures should therefore always be carefully considered to ensure that they are practicable, that compliance can be readily demonstrated, and that it is reasonable to expect employees to carry out the procedures faithfully.'

There is no set pattern of a procedure, but it should clearly spell out the '5 Ws' – What, Where, Who, When, hoW. Typically a procedure is arranged as follows:

1. Purpose             setting out the objectives of the process described in the procedure;
2. Scope               defining the areas of application and exclusions;
3. Person responsible  assigning responsibility to specified person(s);
4. Procedure           listing, step-by-step, what needs to be done and by whom;
5. Records             identifying the records to be generated by applying the procedure.

The following is a collection of experience in writing quality procedures. You may find the hints helpful.

- Break up the process into a sequence of activities in order of time. Follow the chronological sequence in describing the method with particular attention to interfacing the activities.
- Identify the person to take action in each step. Instead of indicating what is to be done, spell out who is to do what.
- Make use of checklists for review and verification.
- Make use of standard forms for recording information.
- Make use of flowcharts instead of writing wherever possible.
- Apply the KISS principle in writing - Keep It Simple and Short.

*120 Writing quality system documents*

A quality procedure should be reviewed and approved before issue and use. This is done by an authorized person, normally the quality manager, whose signature on the document certifies his approval. As the document is subject to document control, the issue number and date of effect must be clearly indicated.

Most quality procedures involve filling in certain forms. These forms, duly completed, become quality records. Design the format of a form so that it is user-friendly, not only to the person who produces the records but also to the person who reads the records. A form does not serve its purpose unless the person using it knows exactly what to put in and how. Wording on the form should be self-explanatory and unambiguous. If confusion is probable, put simple instructions on the back side of the form.

As distinct from a quality (or system) procedure, a technical procedure is a document which details how a particular technical process is to be carried out. Naturally it should include the person(s) responsible, the methods and sequence of work, the criteria or tolerances to be met and the records to be kept. It should also indicate, where appropriate, the equipment and materials to be used, the safety precautions to be taken, and the skill and experience required of the operators.

Whether a particular construction or installation process needs a documented procedure is dependent on answers to the following questions:

- Can every operator perform the work in the same way without being guided by the document?
- Is failure to follow the procedure likely to have adverse effect on the quality of the product?

A technical procedure may be supplemented by one or more work instructions. Whereas the procedure is concerned with the planning and control of a construction or installation process, the work instruction(s) prescribes how the different activities making up the construction process are to be carried out. These subordinate documents provide detailed instructions for the individual operators to perform the tasks assigned to them.

## 8.4 SUMMARY

- There is no particular preference or restriction on the layout of quality system documents, but page identification is essential. The documents

can be in hard copy form or electronic form. Whatever format they are in should facilitate updating.
- The quality manual states the quality policy, elucidates the authorities and responsibilities, and set out the intended quality functions. It includes or makes reference to the quality procedures. A simple way is to provide a list of quality procedures that are bound in a separate volume.
- A procedure is an established method for a specific process. There are two types of procedures: quality (system) procedures and technical procedures. The quality procedures form the basis of the quality system and should cover all applicable requirements of the relevant quality system standard. In writing these documents, remember to address the '5 Ws' – **What, Where, Who, When, hoW.** Also, keep the documents simple and short.

# 9

# Sample quality system documents

## 9.1 OVERVIEW

Documenting a quality system is a time-consuming and painstaking task, especially for the inexperienced. Much effort is spent, and often wasted, in drafting and redrafting the documents. While this is inevitable, it can be minimized if there is an established pattern to follow.

A set of sample documents is appended hereafter to serve as illustrations of the previous chapters. The quality system described is for a *hypothetical* building construction company of medium size with, say, 10 to 20 office staff and a cohort of site staff. The quality system conforms to ISO 9002: 1994. The documentation includes a quality manual and twenty-nine quality procedures, together with forms used in the procedures.

For clarity of illustration, the quality procedures cover separately the twenty system elements of the standard, with one or more procedures written for each element. This arrangement may not be the best for every organization. In writing the quality procedures, the operational processes of the company should remain as entities and the system requirements are built into them. For example, it may be convenient to combine the control of nonconforming product with correct action in a single procedure, but to document the processes of handling, storage and delivery of materials and product separately in different procedures. However, when the quality procedures are process-based as such, they should be carefully checked to ensure that no relevant requirements of the standard are missed out. It is also advisable to provide a listing that references the relevant clauses of the standard against the quality procedures dealing with those clauses.

The quality procedures are for guidance only. In an actual situation, some of these procedures may be irrelevant while additional procedures are needed to describe other processes of the company's operation.

Moreover, the scope of the quality system is often extended to cover safety and environmental control issues.

The quality procedures should not be confused with operational procedures. In a well-established company, there would already be documented procedures for such matters as tendering, project planning, requisition, procurement and personnel management. These and other operational procedures are normally compiled in an operations manual that the quality system documents may make reference to. Besides, there would also be technical procedures, work instructions, checklists, etc. which detail how particular construction, installation and testing processes are to be carried out. They are too numerous to be included here.

The quality procedures make use of a number of forms. These forms, when duly filled in, become quality records. The completed forms should be properly identified, collected, indexed, filed, stored, maintained and ultimately disposed of. In the modern world of information technology, the job is much facilitated by computerization. However, for the sake of clarity and simplicity, the forms are conceived as paper copies in the quality procedures.

The simplistic format of the sample documents may not be aesthetically appealing but it does incorporate all essential features of such documents. This format is adopted to economize printing space of the book. Furthermore, the page of revision record, which is normally included in each quality system document, is left out from the quality procedures, just to avoid duplication of printed material.

It is not at all the intention to establish a model quality system that can be taken off the shelf and applied right away. On the contrary, it is emphasized that a company should establish its own quality system which reflects its business policy, strategies and existing practices. Nevertheless, whoever is commissioned to prepare the documentation will find the sample documents helpful as a sort of template.

This set of sample documents can be curtailed to suit a small business with only a few employees. In such an organization, the proprietor / manager is likely to be the only key person in the management. He/she will play the role of the contracts manager, the construction manager and probably the quality assurance manager as well. The authority is centralized, the communication route shorted, the quality loop reduced, and the procedures are therefore simplified.

In reading this chapter, you are advised to refer to the relevant sections / sub-sections in the previous chapters. Cross-referencing is provided in Table 9.1.

## 124  Sample quality system documents

**Table 9.1** Cross-reference from sample documents to other sections

| Document | Page No. | Related section / sub-section |
|---|---|---|
| Quality manual | 125–50 | 2.4.1, 2.4.2, 4.4 |
| **Quality procedures** | | |
| QP1.1: Review of quality system | 154–5 | 2.4.1, 5.5 |
| QP1.2: Review of project quality management | 156–7 | 2.4.1, 3.4 |
| QP2: Preparation and control of project quality plan | 158–66 | 2.4.2, 3.1, 3.2, 3.3 |
| QP3.1: Tender review | 167–9 | 2.4.3 |
| QP3.2: Contract review | 170–2 | 2.4.3 |
| QP3.3: Variation review | 173–5 | 2.4.3 |
| QP5.1: Control of documents for general application | 176–8 | 2.4.5 |
| QP5.2: Control of documents for specific project | 179–82 | 2.4.5 |
| QP6.1: Evaluation of subcontractors | 183–6 | 2.4.6 |
| QP6.2: Evaluation of suppliers | 187–90 | 2.4.6 |
| QP7: Control of client supplied items | 191–2 | 2.4.7 |
| QP8: Product identification and traceability | 193 | 2.4.8 |
| QP9: Process control | 194–6 | 2.4.9 |
| QP10.1: Receiving inspection and testing | 197–9 | 2.4.10 |
| QP10.2: In-process inspection and testing | 200–2 | 2.4.10 |
| QP10.3: Final inspection and testing | 203–5 | 2.4.10 |
| QP11: Control of measuring and test equipment | 206–8 | 2.4.11 |
| QP12: Inspection and test status | 209 | 2.4.12 |
| QP13.1: Control of nonconforming supply | 210–3 | 2.4.13 |
| QP13.2: Control of nonconforming work | 214–8 | 2.4.13 |
| QP13.3: Handling of client complaints | 219–21 | 2.4.13 |
| QP14.1: Corrective action | 222–5 | 2.4.14 |
| QP14.2: Preventive action | 226–8 | 2.4.14 |
| QP15: Handling, storage and delivery | 229–31 | 2.4.15 |
| QP16: Control of quality records | 232–3 | 2.4.16 |
| QP17: Internal quality audits | 234–40 | 2.4.17, 5.4 |
| QP18.1: Training in quality system | 241 | 2.4.18, 4.5, 5.3 |
| QP18.2: Training in internal quality auditing | 242 | 2.4.18, 4.5 |
| QP18.3: Training in operational / technical skills | 243–6 | 2.4.18 |

*A B C Building*
*Construction Co.*




QUALITY   MANUAL


Issue  1,   27/11/98







Controlled Copy No.
.........................…........

Document Holder
….......................….....

# 𝒜 𝐵 𝐶 𝐵𝑢𝑖𝑙𝑑𝑖𝑛𝑔
Construction Co.




This manual is issued under the authority of the General Manager. It has been reviewed and approved by

*𝒜. 𝑇. 𝐵𝑒𝑠𝑡*
........................
QA Manager

Dated 27/11/98

---

Quality Manual
Issue 1, 27/11/98

Page i of iii
Rev 0

*A B C Building*
*Construction Co.*

RECORD OF REVISIONS

| Revision number | Revision date | Amendment | Pages replaced | Approved by |
|---|---|---|---|---|
| | | | | |
| | | | | |
| | | | | |
| | | | | |
| | | | | |
| | | | | |
| | | | | |
| | | | | |
| | | | | |
| | | | | |

## $\mathcal{A}\,\mathcal{B}\,C$ *Building Construction Co.*

CONTENTS

| | | |
|---|---|---|
| 1 | Foreword | 1 |
| 2 | Quality policy | 2 |
| 3 | Organization structure | 3 |
| 4 | Quality functions | 11 |
| 5 | Project quality management | 21 |

*A B C Building*
*Construction Co.*

1 FOREWORD

This Quality Manual, together with the Quality Procedures it makes reference to, documents the quality system implemented by ABC Building Construction Company in its organization. The quality system has been established to meet the requirements of ISO 9002: 1994. By way of project quality plans, the quality system is applied to all construction projects undertaken by the Company.

This Quality Manual shall be read in conjunction with the Operations Manual of the Company. The two manuals are complementary and cross-referenced.

This Quality Manual is issued under the authority of the General Manager acting as the chief executive of the Company. Distribution and circulation of the document are strictly controlled. The Quality Assurance Manager keeps and maintains a register of controlled copies of the document and their authorized holders.

This Quality Manual will be reviewed as part of the annual management review. Amendments will be issued as required. Amended pages should be properly inserted into the document, and superseded pages removed and returned to the Quality Assurance Manager. Updating of the document should be recorded by filling in the Record of Revisions preceding. Each issue of the document, or amendment thereof, is effective from the date of issue unless otherwise stated.

Words in the masculine gender used in this Quality Manual and the associated documents apply equally to the feminine gender.

*ABC Building
Construction Co.*

## 2 QUALITY POLICY

ABC Building Construction Company is committed to quality of the construction it performs and related services it offers. The objectives of management are to maintain high standard of product while keeping to the client's budget and completion time, and to ensure that all works comply with contractual and regulatory requirements, including safety and environmental requirements.

To achieve the quality objectives, the Company has established a quality system conforming to ISO 9002: 1994. The quality system is fully documented in this Quality Manual and the Quality Procedures it makes reference to.

Quality assurance shall be the concern of staff at all levels. Every section and individual in the Company has quality related responsibilities which shall be honestly and consistently discharged. Should difficulties be encountered, priority is given in using whatever resources available for their speedy resolution.

All personnel in the Company shall abide by this policy and fully cooperate with the Quality Assurance Manager in the implementation and maintenance of the quality system.

*J. E. Prosper*
..........................
General Manager

Dated 18/7/98

*A B C Building
Construction Co.*

## 3  ORGANIZATION STRUCTURE

ABC Building Construction Company is governed by the Board of Directors. The executive power is vested in the General Manager who is the chief executive officer of the Company. The General Manager is assisted by the managers in the Company's operations.

The Quality Assurance Manager is the management representative on operational matters related to quality. He has specific authority delegated by the General Manager to implement and maintain the quality system. He is functionally independent of other managers and reports directly to the General Manager.

The section heads are responsible for identifying and providing the required manpower and equipment for efficient operation in their respective sections. They are also responsible for providing appropriate training to their staff and workers before assigning them to duties which require special skills.

The organization chart shows the interrelation of personnel who manage, perform and verify work affecting quality. The site organization structure is indicative only, and may vary from project to project. The position descriptions of key members of management are shown on the following pages. Those of other staff are included in the individual personal files.

For each project, a project manager is assigned to be in charge. A project manager may handle more than one project if the total volume of work is not excessive, in which case a site agent is stationed on each site to act as his/her deputy. A team of site staff commensurate with the size and complexity of the project is allocated from the existing establishment or specially recruited for the project.

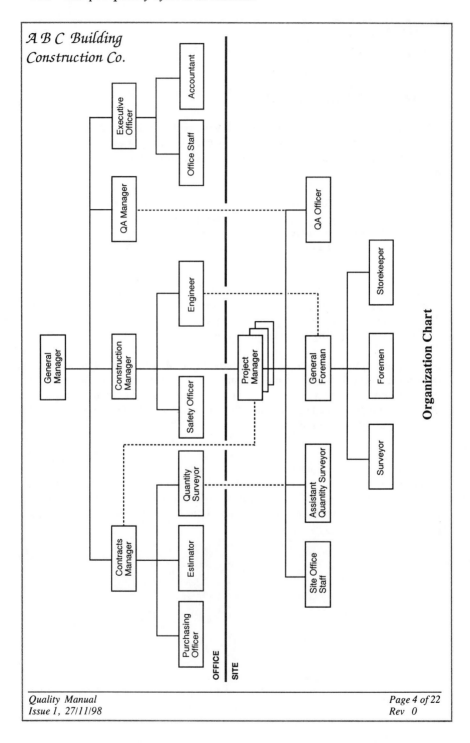

# A B C Building Construction Co.

POSITION :       General Manager

SECTION :        Administration

REPORTING TO :   Board of Directors

RESPONSIBILITY & AUTHORITY :

1  To direct and oversee the overall operation of the Company

2  To negotiate and arrange finance for a project

3  To submit a tender, and to enter into a contract, on behalf of the Company

4  To set the quality policy and objectives

5  To review the quality system periodically with other managers

6  To take overall responsibility for the quality of all works and services undertaken by the Company

*A B C Building*
*Construction Co.*

POSITION : Construction Manager

SECTION : Construction

REPORTING TO : General Manager

RESPONSIBILITY & AUTHORITY :

1  To direct and oversee construction operations, including operational strategy, equipment and manpower deployment

2  To advise the General Manager regularly on the current capabilities of the Company for new commitments

3  To advise the General Manager regularly on the current and projected needs of technical resources of the Company

4  To assign technical staff with appropriate qualifications and experience to a project

5  To provide resources for a project to meet the construction programme

6  To identify training needs of technical staff and to provide resources for necessary training

## A B C Building Construction Co.

POSITION :       Contracts Manager

SECTION :        Contracts

REPORTING TO :   General Manager

RESPONSIBILITY & AUTHORITY :

1  To direct and oversee the execution of contracts

2  To work out an estimate for a new project and to prepare a tender

3  To exercise cost control of ongoing projects

4  To settle disputes related to a contract

5  To authorize major purchasing functions, including plant hire and material supply

6  To select subcontractors and to enter into subcontracts on behalf of the Company

# ABC Building Construction Co.

POSITION :        Quality Assurance Manager

SECTION :         Quality Assurance

REPORTING TO :    General Manager

RESPONSIBILITY & AUTHORITY :

1. To publicize the quality policy so that it is understood at all levels of the organization

2. To work with other section heads in implementing and maintaining the quality system at all levels of the organization

3. To report on the performance of the quality system to the management for review

4. To establish and maintain control of documents relating to the quality system, including review and approval of such documents

5. To make amendments to quality procedures prompted by quality deficiencies, nonconformance or client complaints

6. To direct and oversee collection, storage and disposal of quality records

7. To plan, schedule and arrange internal quality audits

8. To organize and conduct quality related training, including training of internal quality auditors

9. To liaise with the certification body and other external parties on matters relating to the quality system

10. To assist the Project Manager in the preparation of project quality plans

# A B C Building Construction Co.

POSITION :        Executive Officer

SECTION :         Administration

REPORTING TO :    General Manager

RESPONSIBILITY & AUTHORITY :

1   To direct and oversee office administration

2   To organize recruitment, training and upgrading of staff

3   To keep and maintain staff records

4   To distribute, retrieval and dispose of controlled copies of documents

5   To supervise accounts and cash flow

6   To liaise with clients, bankers and auditors relating to financial matters

## ABC Building Construction Co.

POSITION :	Project Manager

SECTION :	Construction

REPORTING TO :	Construction Manager

RESPONSIBILITY & AUTHORITY :

1. To prepare a construction programme for the project he/she is in charge of, including site organization and resource scheduling

2. To prepare a quality plan for the project

3. To coordinate the day-to-day activities on site

4. To schedule purchase and delivery of materials and equipment

5. To coordinate subcontracted work and to exercise control on subcontractors

6. To monitor the progress of the project

7. To deal with client complaints and to resolve disputes

8. To review, and to authorize the disposition of, nonconforming work

9. To take corrective and preventive action prompted by quality deficiencies on site

*A B C Building
Construction Co.*

## 4 QUALITY FUNCTIONS

### 4.1 Management review

The General Manager, together with the Quality Assurance Manager and other section heads, reviews the quality system once a year, normally by the end of June. Additional review will be performed whenever the General Manager considers it necessary. The corporate management review is conducted and recorded in accordance with Quality Procedure QP1.1. In the review, the continuing suitability and effectiveness of the quality system are assessed in the light of the results of the quality audits, amendments in government regulations and clients' needs. Appropriate changes are made to the quality system if necessary. Effects of changes initiated by the previous review are also assessed. Results of the corporate management review are reported to the Board of Directors.

For an individual project, the Project Manager carries out a project management review once in six months and whenever he finds an urgent need to do so. The review is conducted and recorded in accordance with Quality Procedure QP1.2. The project quality plan is updated and reissued if necessary. Results of the project management review are forwarded to the Quality Assurance Manager for inclusion on the agenda of the corporate management review when it is next held.

### 4.2 Quality system

The Company implements and maintains a quality system conforming to ISO 9002:1994. Documentation of the quality system comprises this Quality Manual and the Quality Procedures referred to by this Manual and listed in Table 1. Standard forms are used in connection with some procedures. Work instructions related to a procedure are prepared whenever and wherever necessary.

For an individual project, quality planning, as described in Section 5 of this Manual, is undertaken as soon as practical. A quality plan is prepared in accordance with Quality Procedure QP2.

## A B C Building Construction Co.

**Table 1** List of quality procedures

| Document No. | Title of quality procedure | Relevant clause in ISO 9002:1994 |
|---|---|---|
| QP1.1 | Review of quality system | 4.1 |
| QP1.2 | Review of project quality management | 4.1 |
| QP2 | Preparation and control of project quality plan | 4.2 |
| QP3.1 | Tender review | 4.3 |
| QP3.2 | Contract review | 4.3 |
| QP3.3 | Variation review | 4.3 |
| QP4 | Void | |
| QP5.1 | Control of documents for general application | 4.5 |
| QP5.2 | Control of documents for specific projects | 4.5 |
| QP6.1 | Evaluation of subcontractors | 4.6 |
| QP6.2 | Evaluation of suppliers | 4.6 |
| QP7 | Control of client-supplied items | 4.7 |
| QP8 | Product identification and traceability | 4.8 |
| QP9 | Process control | 4.9 |
| QP10.1 | Receiving inspection and testing | 4.10 |
| QP10.2 | In-process inspection and testing | 4.10 |
| QP10.3 | Final inspection and testing | 4.10 |
| QP11 | Control of measuring and test equipment | 4.11 |
| QP12 | Identification of inspection and test status | 4.12 |
| QP13.1 | Control of nonconforming supply | 4.13 |
| QF13.2 | Control of nonconforming work | 4.13 |
| QP13.3 | Handling of client complaints | 4.13 |
| QP14.1 | Corrective action | 4.14 |
| QP14.2 | Preventive action | 4.14 |
| QP15 | Handling, storage and delivery | 4.15 |
| QP16 | Control of quality records | 4.16 |
| QP17 | Internal quality audits | 4.17 |
| QP18.1 | Training in quality system | 4.18 |
| QP18.2 | Training in quality auditing | 4.18 |
| QP18.3 | Training in operational / technical skills | 4.18 |
| QP19 | Void | |
| QP20 | Void | |

*Quality Manual*
*Issue 1, 27/11/98*

*A B C Building*
*Construction Co.*

### 4.3 Tender, contract and variation review

An invitation to tender for a new project is deliberated and a tender prepared, if so decided, in accordance with the Operations Manual.

The prepared tender is reviewed in accordance with Quality Procedure QP3.1 before submission. This is to ensure that
  (a) the client's requirements are adequately defined and documented in the tender; and
  (b) the Company has the capability to meet the requirements.

If the tender is successful, the contract offered by the client is reviewed in accordance with Quality Procedure QP3.2 before acceptance. This is to ensure that any differences between the contract requirements and those in the tender are resolved.

During execution of the contract, any variation order involving significant changes of requirements either in kind or in quantity or both is reviewed in accordance with Quality Procedure QP3.3 before acceptance. This is to ensure that
  (a) the amended requirements are adequately defined and documented in the variation order and accompanying drawings; and
  (b) the amended requirements can be satisfactorily accommodated.

Changes in requirements are transferred to functions concerned within the Company through control of contract documents as per Quality Procedure QP5.2.

### 4.4 Design control

The Company's operations do not include design of permanent works. (Design of temporary works is part of the construction process and is subject to process control.)

### 4.5 Document and data control

All documents describing the Company's operations are reviewed and approved for adequacy by the Quality Assurance Manager prior to issue. Any revised document goes through the same process. A master list is established identifying the current revision status of the documents.

*ABC Building*
*Construction Co.*

Distribution and retrieval of documents are controlled by the Executive Officer following Quality Procedure QP5.1. Controlled copies are distributed to the section heads who will place them at such locations as to be easily accessible. A controlled copy is also kept in each site office. Obsolete documents are promptly removed and returned to the Executive Officer for disposal.

Distribution and retrieval of project-specific documents are controlled by the Contracts Manager / Project Manager following Quality Procedure QP5.2.

**4.6 Subcontracting and purchasing**

A list of acceptable subcontractors is established and maintained for each trade. No subcontract is offered to anyone not on the list, except with the special permission of the General Manager. Where the contract calls for subcontractors with a quality system certified to a national or international standard, only those so qualified are employed.

Potential subcontractors are subject to evaluation before inclusion in the appropriate list. Subcontractors currently employed by the Company are re-evaluated annually and at completion of each subcontract. The process of evaluation and re-evaluation of subcontractors follows Quality Procedure QP6.1.

A subcontract must contain adequate information showing, inter alia, the quality of service expected and the quality assurance required. The subcontract is reviewed and approved for adequacy of the specified requirements prior to release. Preparation of subcontracts, selection of subcontractors, with or without tendering, and award of subcontracts are governed by the Operations Manual.

A list of acceptable suppliers is established and maintained for each major item of material. The process of evaluation and re-evaluation of suppliers follows Quality Procedure QP6.2.

A purchase order must contain data clearly describing the material or equipment ordered, including where applicable the type, class, grade or other precise identification, and the applicable specification or national standard. The purchase order is reviewed and approved for adequacy of the specified requirements prior to release. Requisition for material / equipment supply, calling for quotations and issue of purchase order are governed by the Operations Manual.

## ABC Building Construction Co.

If verification at source is a contract requirement, provisions are made in the subcontracts and/or purchase orders to afford the client or his representative the right to do so at the subcontractor's and/or supplier's premises. Notwithstanding such verification, receiving inspection and other quality control measures are conducted as indicated in the project quality plan.

### 4.7 Control of client supplied items

Items supplied by the client for incorporation into the permanent works or for related activities are inspected, stored and used in accordance with Quality Procedure QP7. A record is maintained indicating the type, quantity, condition and verification status of each item. Any item that is lost, damaged or is otherwise unsuitable for use is recorded and reported to the client.

### 4.8 Product identification and traceability

Precast or prefabricated units and sub-assemblies are so marked as to relate to the contract drawings. Where traceability is specified in the contract, each individual unit is given a unique mark and its final location in the structure is recorded. Product identification and traceability is accomplished by following Quality Procedure QP8.

Traceability of materials used and appliances installed, if required, is accomplished by controlling the receipt and issuance of the items as per Quality Procedure QP10.1.

### 4.9 Process control

Before construction work commences on a project, a quality plan is prepared in accordance with Quality Procedure QP2. The quality plan indicates the construction processes involved and the construction schedule, the materials, labour and equipment required and their procurement, the training needs for special processes, the procedures or work instructions to be followed, and the records to be established.

Based on the construction schedule, arrangements are made for the provision of skilled labour, suitable equipment and regular supervision at the appropriate time. Procedures or work instructions are prepared in advance for non-routine or unconventional processes. Progress of work is monitored as per Quality Procedure QP9 and recorded on a bi-weekly programme.

## ABC Building Construction Co.

Design of temporary works, unless subcontracted to a technical consultant, is performed by the Engineer. The design is checked and approved by the Construction Manager before being implemented.

### 4.10 Inspection and testing

Materials, components, and appliances received on site, except minor items of small quantities, are subject to receiving inspection and testing in accordance with Quality Procedure QP10.1. The amount and nature of receiving inspection and testing are indicated in the project quality plan. No incoming item is used or installed until it has been verified as conforming to specified requirements. Where an incoming item is needed for urgent use or installation prior to verification, it can only be released with approval of the Project Manager and the location where it is used or installed is recorded.

Construction work in progress is inspected and/or tested in accordance with Quality Procedure QP10.2. The nature of in-process inspection and testing and the various stages when it is performed are indicated in the inspection and test plan which forms part of the project quality plan. The construction process is not allowed to continue until the inspector is satisfied with the quality of work completed so far.

Any finished construction work is subject to final inspection and testing in accordance with Quality Procedure QP10.3. The nature of final inspection and testing is indicated in the inspection and test plan. This includes verification that all in-process inspection and tests have been carried out and that the results meet specified requirements.

No finished work can be covered over or built upon until the final inspection and testing has been passed. However, products of a material which takes time to develop its full strength (e.g. concrete) may be provisionally accepted pending satisfactory test results.

Records are kept of all inspections and/or tests performed, showing in each case whether the material or work has passed or failed the inspection and/or test according to defined acceptance criteria. Any nonconforming material is dealt with in accordance with Quality Procedure QP13.1 and any nonconforming work in accordance with Quality Procedure QP13.2.

*A B C Building*
*Construction Co.*

### 4.11 Control of measuring and test equipment

Measuring and test equipment is controlled, maintained and calibrated in accordance with Quality Procedure QP11. Reports of calibration are kept on site where the equipment is currently being used. When a piece of equipment is found to be out of calibration, the validity of previous measurements made with it is assessed and documented.

Where close tolerance is specified in the contract for certain measurement, only the equipment capable of the necessary accuracy and precision will be used for the particular measurement. Such equipment is indicated in the project quality plan.

### 4.12 Inspection and test status

The inspection and test status of a consignment of material, components or appliances is indicated on the 'Material Receipt and Issue' form for the consignment. Any defective items are identified and separately stored.

The inspection and test status of precast units and other prefabricated components is indicated by marking as per Quality Procedure QP12.

The inspection and test status of finished work is indicated by signing off the inspection and test plan for the work.

### 4.13 Control of nonconforming supply and nonconforming work, and handling of client complaint

A material, component or appliance which does not pass the receiving inspection and testing is identified and segregated. The nonconforming item is reviewed and dispositioned in accordance with Quality Procedure QP13.1.

Finished or semi-finished work which does not conform to specified requirements is marked and, wherever possible, segregated. The nonconforming work is reviewed and dispositioned in accordance with Quality Procedure QP13.2. If the nonconforming work is repaired or re-built, it will be inspected or tested again before acceptance.

*A B C Building*
*Construction Co.*

A complaint of the client or his representative about the construction in progress or execution of the contract is reviewed and settled in accordance with Quality Procedure QP13.3.

Separate registers are established for nonconforming supply, nonconforming work and client complaints in order to detect any trend of recurrence of nonconformance.

### 4.14 Corrective and preventive action

The cause of an incident of nonconformance or client complaint is investigated and the results of investigation are recorded. Appropriate corrective and preventive action is taken to eliminate the cause(s) of actual or potential problems. The investigation and the subsequent actions are carried out in accordance with Quality Procedure QP14. A register of corrective and preventive action is established and maintained. Changes made to quality procedures and/or work instructions are subject to document control as per Quality Procedure QP5.1.

### 4.15 Handling, storage and delivery

Materials for construction and appliances for installation are handled and stored in such a way as to prevent damage or deterioration. Adequate and appropriate areas on site are designated for storage purposes as marked on the site layout plan. Receipt to and dispatch from such areas are authorized and recorded in a 'Material Receipt and Issue' form as per Quality Procedure 10.1. Materials that are liable to deteriorate with time are inspected at regular intervals. Any material that is found to be unsuitable for use is clearly identified or immediately removed from the storage area.

Precast and prefabricated units produced on site or off site are handled, stored and delivered in accordance with Quality Procedure QP15. Receipt to and dispatch from the storage area are authorized by signing an entry in the inventory of such units.

*A B C Building*
*Construction Co.*

### 4.16 Control of quality records

Quality records, including pertinent records from the subcontractors, are maintained to demonstrate conformance to specified requirements and effective operation of the quality system. All quality records are identified, collected, indexed, filed, stored, accessed and disposed of in accordance with Quality Procedure QP16. Project-specific records are kept for seven years after completion of the contract unless otherwise stated in the contract. Other records are kept for two years or as directed by the Quality Assurance Manager.

Where specified in the contract, quality records are made available for evaluation by the client or his representative for the period stated.

### 4.17 Internal quality audits

Quality activities and related results are audited to verify whether the quality procedures are followed and the quality plans are implemented. The audits are carried out by personnel who have been so trained and are independent of those having direct responsibility for the activity being audited. Each section of the head office and each construction site is audited at least once in a year, the frequency of audit being based on the status and importance of the activity to be audited.

Internal quality audits are scheduled, planned, conducted and recorded in accordance with Quality Procedure QP17. Results of the audits are brought to the attention of the personnel having responsibility in the area audited. Corrective action is taken, if necessary, in accordance with Quality Procedure QP14. Depending on the severity of nonconformance found during the audit, a follow-up audit may be conducted to verify that corrective action has been taken and it is effective.

### 4.18 Training

The Company ensures that staff at all levels understand the quality policy and the quality procedures related to their work. The Company also ensures that those being assigned to specific tasks are qualified and/or adequately trained. Training in quality system, quality auditing and technical skills is provided and recorded in accordance with Quality Procedures QP18.1, QP18.2 and QP18.3 respectively.

*A B C Building*
*Construction Co.*

**4.19 Servicing**

The Company does not undertake regular maintenance or servicing of finished building works. (Defects discovered during the defects liability period are treated in the same manner as client complaints.)

**4.20 Statistical techniques**

No statistical techniques are used for establishing, controlling and verifying process capability and product characteristics.

*A B C Building*
*Construction Co.*

## 5 PROJECT QUALITY MANAGEMENT

As soon as the Company is commissioned to carry out a project, a project team is formed and specific responsibility and authority assigned to each member of the team. The project team is briefed on the project requirements by the personnel who prepared the tender. If certain member(s) of the team require(s) training, it will be provided in time for the job.

The Project Manager, together with the Construction Manager and other staff, works out a construction programme (and sub-programmes if appropriate) for the project. In the process, any special equipment needed is identified and arrangement made for its timely purchase or hiring so that construction can go ahead without interruption or delay. Requirements for special skills are also identified and provided for by training existing staff or recruiting new staff.

Subcontractors, and suppliers of major items of material, are selected from the current lists of acceptable subcontractors and suppliers with or without tendering. Where specified in the contract, only those subcontractors and suppliers who have a certified quality system in place are appointed.

The project team, under the direction of the Project Manager, considers the verification requirements of materials and, if necessary, requests the Contracts Manager to commission a testing agency in due course to carry out sample testing. The project team also considers the inspection and testing needs at various stages of construction, and identifies any testing equipment and skills to be acquired. Inspection and test plans are prepared, both for work by direct labour and for subcontracted work. Alternatively certain subcontractors are each required to submit an inspection and test plan for the part of work he undertakes to perform.

The Quality Assurance Manager prepares a schedule of on-site quality audits based on the anticipated progress of the project, and makes provisions for conducting the audits in due course. He also arranges for an audit of the subcontractor's quality system (to the extent that it affects the quality of the subcontracted work), should this be a condition of the subcontract.

Based on the outcome of quality planning, the Project Manager and the Quality Assurance Manager jointly prepare a quality plan, following Quality Procedure QP2. The quality plan makes reference to the documented procedures

*A B C Building
Construction Co.*

of the Company's quality system, with due modifications to cater for project-specific requirements. Where the project involves special processes, additional procedures and/or work instructions are drafted. The quality plan is to be ready as soon as possible, but not later than commencement of site work. It is submitted to the client for information, and for approval if required by contract, before implementation. It is updated regularly as the project proceeds, normally every three months.

Quality management of the project is reviewed at regular intervals, at least once every six months. The review is conducted and recorded as per Quality Procedure QP1.2. The quality plan is revised if necessary.

*A B C Building*
*Construction Co.*

# QUALITY PROCEDURES
## ( WITH FORMS )

Controlled Copy No.  …....................….........

Document Holder  …..........................…....

## A B C Building Construction Co.

LIST OF QUALITY PROCEDURES

| Document No. | Procedure title | Issue date | | |
|---|---|---|---|---|
| | | Issue 1 | Issue 2 | Issue 3 |
| QP1.1 | Review of quality system | 27/11/98 | | |
| QP1.2 | Review of project quality management | 27/11/98 | | |
| QP2 | Preparation and control of project quality plan | 27/11/98 | | |
| QP3.1 | Tender review | 27/11/98 | | |
| QP3.2 | Contract review | 27/11/98 | | |
| QP3.3 | Variation review | 27/11/98 | | |
| QP5.1 | Control of documents for general application | 27/11/98 | | |
| QP5.2 | Control of documents for specific project | 27/11/98 | | |
| QP6.1 | Evaluation of subcontractors | 27/11/98 | | |
| QP6.2 | Evaluation of suppliers | 27/11/98 | | |
| QP7 | Control of client supplied items | 27/11/98 | | |
| QP8 | Product identification and traceability | 27/11/98 | | |
| QP9 | Process control | 27/11/98 | | |
| QP10.1 | Receiving inspection and testing | 27/11/98 | | |
| QP10.2 | In-process inspection and testing | 27/11/98 | | |
| QP10.3 | Final inspection and testing | 27/11/98 | | |
| QP11 | Control of measuring and test equipment | 27/11/98 | | |
| QP12 | Inspection and test status | 27/11/98 | | |

*A B C Building Construction Co.*

LIST OF QUALITY PROCEDURES (contd)

| Document No. | Procedure title | Issue date | | |
|---|---|---|---|---|
| | | Issue 1 | Issue 2 | Issue 3 |
| QP13.1 | Control of nonconforming supply | 27/11/98 | | |
| QP13.2 | Control of nonconforming work | 27/11/98 | | |
| QP13.3 | Handling of client complaints | 27/11/98 | | |
| QP14.1 | Corrective action | 27/11/98 | | |
| QP14.2 | Preventive action | 27/11/98 | | |
| QP15 | Handling, storage and delivery | 27/11/98 | | |
| QP16 | Control of quality records | 27/11/98 | | |
| QP17 | Internal quality audits | 27/11/98 | | |
| QP18.1 | Training in quality system | 27/11/98 | | |
| QP18.2 | Training in internal quality auditing | 27/11/98 | | |
| QP18.3 | Training in operational / technical skills | 27/11/98 | | |

*A B C Building*
*Construction Co.*

## QP 1.1  PROCEDURE FOR REVIEW OF QUALITY SYSTEM

**1  Purpose**
To specify actions and to assign responsibilities for scheduling, conducting and recording management review

**2  Scope**
Applicable to annual review of the quality system

**3  Person responsible**
Quality Assurance Manager

**4  Procedure**

4.1 The Quality Assurance Manager, in consultation with the General Manager, fixes a date for an annual meeting to review the quality system. The meeting is normally held by the end of September and is to be attended by the following persons:
- General Manager            (Chairperson)
- Quality Assurance Manager  (Secretary)
- Construction Manager
- Contracts Manager
- Executive Officer
- Other staff by invitation

4.2 The QA Manager requests each section head to review the quality procedures applicable to his area of responsibility in the light of results of internal quality audits, and to submit a report not later than one week before the date of the meeting.

4.3 The QA Manager prepares the agenda of the meeting and distributes it to the participants at least three working days before the meeting. The following items are normally on the agenda, and anyone may suggest additional items to be included.
- reports of quality audits, both internal and external
- nonconforming work discovered and client complaints received since last review
- corrective / preventive actions taken since last review
- changes in quality procedures since last review
- review, and revision if necessary, of quality manual
- minutes of project management review meetings

*A B C Building*
*Construction Co.*

- training needs arising from staff changes and/or quality deficiencies
- changes in statutory regulations, if any, and their impact on quality management
- changes in Company's business direction, if any, and its impact on quality management

4.4 The General Manager and the QA Manager must be present at the meeting. Any other person who is unable to attend must send a representative to the meeting equipped with full instructions and relevant documents.

4.5 The QA manager takes minutes as a formal record of the meeting which should indicate the follow-up actions and the persons responsible.

4.6 The General Manager forwards the confirmed minutes to the Board of Directors when it is next held.

4.7 The QA Manager amends the quality system documents as decided at the meeting.

5 **Records**
Minutes of review meeting

*A. T. Best*
Procedure approved by ........................
*QA Manager*
Dated 27/11/98

*A B C Building*
*Construction Co.*

## QP 1.2 PROCEDURE FOR REVIEW OF PROJECT QUALITY MANAGEMENT

**1 Purpose**
To specify actions and to assign responsibilities for scheduling, conducting and recording management review on site

**2 Scope**
Applicable to periodic review of quality management of a project

**3 Person responsible**
Project Manager

**4 Procedure**

4.1 The Project Manager fixes a date for a review meeting which should be not later than six months after the last review. The meeting is to be attended by the following persons:
- project Manager (Chairperson)
- quality Assurance Officer (Secretary)
- construction Manager or his deputy
- general Foreman
- other site staff by invitation
- subcontractor(s) by invitation

4.2 Under the instructions of the Project Manager, the Quality Assurance Officer prepares the agenda of the meeting and distributes it to the participants at least three working days before the meeting. The following items are normally on the agenda, and anyone may suggest additional items to be included.
- progress of project
- changes in construction programme and/or resource provisions
- changes in contract requirements, including variation orders
- changes in applicable statutory regulations and their impact
- nonconformances and client complaints arising since last review and corrective/preventive actions taken
- performance of subcontractors and suppliers
- training needs arising from staff changes or quality deficiencies

4.3 The QA Officer takes minutes as a formal record of the meeting which should indicate the follow-up actions and the persons responsible.

*A B C Building*
*Construction Co.*

4.4 The Project Manager forwards the confirmed minutes of the meeting to the QA Manager.

4.5 The Project Manager amends the quality plan as decided at the meeting and reissues the revised document, or pages thereof, as per Quality Procedure QP 5.2.

**5   Records**
Minutes of review meeting

Procedure approved by  *A. T. Best*
........................
QA Manager

Dated  27/11/98

*A B C Building*
*Construction Co.*

**QP 2  PROCEDURE FOR PREPARATION AND CONTROL OF PROJECT QUALITY PLAN**

**1  Purpose**
To specify actions for preparation and control of a quality plan for a project

**2  Scope**
Applicable to all construction projects undertaken by the Company on contract basis

**3  Person responsible**
Project Manager

**4  Procedure**

4.1  The Project Manager prepares a site organization chart, with named personnel if known, and specifies the responsibility and authority of each member of the project team. Should training need be identified in the process, the Project Manager notifies the Executive Officer who will assist him in providing the person(s) concerned with appropriate training in time for the job, as per Quality Procedure QP18.3.

4.2  The Project Manager, in collaboration with the Construction Manager, prepares a construction programme (and sub-programmes if appropriate) for the project. Due consideration is given to specific requirements of the project such as:
- possession of site in stages or in parts
- hand-over of works in stages or in parts
- environmental control, eg. traffic restriction, noise abatement

4.3  The Project Manager prepares a schedule of subcontracting based on the construction programme, using Form QF2-1.

4.4  The Project Manager prepares a schedule of material procurement based on the construction programme, using Form QF2-2. Only major items of materials and appliances are included.

4.5  The Project Manager prepares a schedule of equipment procurement based on the construction programme, using Form QF2-3.

### A B C Building
### Construction Co.

4.6 The Project Manager, assisted by the project team, extracts the verification requirements of materials from the contract documents and relevant standards, and makes a summary of the requirements using Form QF2-4. If possible, items are listed separately as follows:
   (i) items that are supplied with manufacturer's test certificate
   (ii) items that require sample approval by Architect/Consulting Engineer
   (iii) items that require sample testing by test agency

4.7 The Project Manager, assisted by the project team, identifies the construction and installation processes for which inspection and test plans (ITP) are required. If any ITP is not already available, he directs the project team to prepare it in time for use. Form QF2-5 is used whenever suitable.

*Note : The Project Manager may require a subcontractor to prepare an ITP for the work he undertakes to perform and submit it for approval and, if appropriate, incorporation in the quality plan.*

4.8 The Project Manager lists the quality procedures and work instructions in the Company quality system that are applicable to the project and, with the assistance of the Quality Assurance Manager, makes necessary modifications to suit the specific requirements of the project. He also prepares procedures and/or work instructions for special processes before they are carried out.

4.9 The Project Manager and the Quality Assurance Manager jointly decide on the frequency of internal quality audit of the project, including audit of subcontractors if it is deemed necessary. Provisional dates are set if possible.

4.10 The Project Manager identifies the quality records to be retained, including pertinent quality records from the subcontractors.

4.11 The Project Manager, in collaboration with the Quality Assurance Manager, compiles the information collected above into a quality plan. The quality plan typically contains the following:
   - title page showing project title and contract number, document authorization, issue number and date of effect
   - record of revisions
   - brief description of project
   - list of contract documents and drawings

*Quality Procedure QP2*
*Issue 1, 27/11/98*

*A B C Building*
*Construction Co.*

- quality objectives of project
- project organization chart, with named personnel if known
- responsibilities and authorities of project staff
- site layout plan
- construction programme and sub-programmes
- schedule of subcontracting
- schedule of material procurement
- schedule of equipment procurement
- verification requirements of materials
- inspection and test plans, or list thereof
- list of quality procedures and work instructions (in Company quality system) applicable to project
- list of project-specific procedures and work instructions, including modified quality procedures (If a procedure or work instruction is not ready when the quality plan is issued, a target date for its provision is indicated.)
- list of quality records to be retained
- frequency (or provisional dates) of internal quality audit, including audit of subcontractors if necessary
- frequency of updating the quality plan

4.12 The Project Manager circulates the draft quality plan to the Contracts Manager and the Construction Manager for comments and amendments, and then submits it to the Architect / Consulting Engineer for information, and for approval if required by contract.

4.13 The Project Manager reviews and approves the finalized quality plan before release for use.

4.14 On site, the Quality Assurance Officer controls distribution of the quality plan in accordance with Quality Procedure QP5.2. He keeps a controlled copy in the site office for reference. Under the direction of the Project Manager, he issues controlled copies to those persons who need them for their operation.

*Note : The construction programme and sub-programmes, as well as the inspection and test plans, can be issued as separate documents subject to document control. Controlled copies of these documents are distributed to relevant subcontractors and suppliers when required.*

*Quality Procedure QP2*
*Issue 1, 27/11/98*

*A B C Building*
*Construction Co.*

**5  Records**
Project quality plan
Form QF2-1  Schedule of subcontracting
Form QF2-2  Schedule of material procurement
Form QF2-3  Schedule of equipment procurement
Form QF2-4  Verification requirements of materials
Form QF2-5  Inspection and test plan

**6  Appendix**
Processes for which ITPs are normally required include, but are not limited to, the following:
- excavation and earthwork
- piling and caisson work
- precasting
- concrete work, including falsework, formwork and prestressing
- structural steelwork
- cladding, facade and curtain walling
- waterproofing
- plumbing and drainage
- mechanical services
- electrical services
- fire-fighting facilities

For most of these processes, ITPs are already available, such as from previous projects. They may be adopted with or without modification.

Procedure approved by  *A. T. Best*
                      ........................
                      QA Manager

Dated  27/11/98

*A B C Building*
*Construction Co.*

## SCHEDULE OF SUBCONTRACTING

Project :

Contract No. :

| Trade / Service | Date of nomination | Date of start work | Subcontract No. | Name of subcontractor |
|---|---|---|---|---|
|  |  |  |  |  |
|  |  |  |  |  |
|  |  |  |  |  |
|  |  |  |  |  |
|  |  |  |  |  |
|  |  |  |  |  |
|  |  |  |  |  |
|  |  |  |  |  |
|  |  |  |  |  |
|  |  |  |  |  |
|  |  |  |  |  |
|  |  |  |  |  |

Note : If a date is provisional, put it in brackets.

*Form QF2-1*
*Issue 1, 27/11/98*

# A B C Building Construction Co.

## SCHEDULE OF MATERIAL PROCUREMENT

Project :

Contract No. :

| Material / Appliance | Date of purchase | Date of first delivery | Purchase order No. | Name of supplier |
|---|---|---|---|---|
|  |  |  |  |  |
|  |  |  |  |  |
|  |  |  |  |  |
|  |  |  |  |  |
|  |  |  |  |  |
|  |  |  |  |  |
|  |  |  |  |  |
|  |  |  |  |  |
|  |  |  |  |  |
|  |  |  |  |  |
|  |  |  |  |  |
|  |  |  |  |  |

Note : If a date is provisional, put it in brackets.

*Form QF2-2*
*Issue 1, 27/11/98*

*A B C Building*
*Construction Co.*

SCHEDULE OF EQUIPMENT PROCUREMENT

Project :

Contract No. :

| Equipment / Plant | Model / Capacity | Accuracy req'd, if unusual | Period required | Expected source |
|---|---|---|---|---|
|  |  |  |  |  |
|  |  |  |  |  |
|  |  |  |  |  |
|  |  |  |  |  |
|  |  |  |  |  |
|  |  |  |  |  |
|  |  |  |  |  |
|  |  |  |  |  |
|  |  |  |  |  |
|  |  |  |  |  |
|  |  |  |  |  |
|  |  |  |  |  |

Notes :  1. Equipment includes surveying instruments and testing devices.
2. Source could be 'hiring', 'purchase', 'from another site', 'provided by subcontractor', etc.

*Form QF2-3*
*Issue 1, 27/11/98*

# $\mathcal{ABC}$ $\mathcal{B}uilding$ $Construction\ Co.$

## VERIFICATION REQUIREMENTS OF MATERIALS

Project :

Contract No. :

| Material / Appliance | Supplier's test certificate required ? | Approval of sample required ? | Expected date of sample submission | Sample testing requirement ||||
|---|---|---|---|---|---|---|---|
| | | | | Test(s) required* | Test frequency | Acceptance criteria | Test record |
| | | | | | | | |
| | | | | | | | |
| | | | | | | | |
| | | | | | | | |
| | | | | | | | |
| | | | | | | | |
| | | | | | | | |
| | | | | | | | |

\*  *If no test is required, enter 'N/A'.*

*Form QF2-4*
*Issue 1, 27/11/98*

# ABC Building Construction Co.

## INSPECTION AND TEST PLAN

Project :
Contract No. :
Subcontract :

ITP No. :
Revision No. & date :
Work covered :

| Activity | Inspection / Test | Acceptance criteria | Records / Remarks | Verifying party ||| Sign & date |
|---|---|---|---|---|---|---|---|
| | | | | SC | C | A/E | RA |
| | | | | | | | |
| | | | | | | | |
| | | | | | | | |
| | | | | | | | |
| | | | | | | | |
| | | | | | | | |

Verification code :  I   Inspection
                         T   Test
                         W   Witness
                         H   Hold
                         D   Document review

Legend :   SC   Subcontractor
              C     Contractor
              A/E   Architect/Engineer
              RA    Regulatory authority

*Form QF2-5*
*Issue 1, 27/11/98*

*A B C Building*
*Construction Co.*

**QP 3.1   PROCEDURE FOR TENDER REVIEW**

**1   Purpose**
  (i)   To ensure that the project requirements are adequately defined and documented
  (ii)  To assess the Company's capability to undertake the project

**2   Scope**
Applicable to tenders for new projects

**3   Person responsible**
Contracts Manager

**4   Procedure**

4.1   The Contracts Manager, in collaboration with the Construction Manager, appraises the scope of work, the construction method, construction period and resource implications, and projects the Company's capability to undertake the work in addition to existing commitments.

4.2   The Contracts Manager and his staff study the tender documents, noting the requirements and quantities. Should any requirement be inadequately defined or documented, he seeks clarification from the Architect / Consulting Engineer.

4.3   The Contracts Manager and his staff prepare a tender for the project, following the guidelines in the Operations Manual.

4.4   The Contracts Manager completes the checklist on Form QF3-1.

4.5   The Contracts Manager arranges a meeting with the General Manager and the Construction Manager to review the prepared tender and to resolve any problem/uncertainty indicated on the checklist. The review meeting is recorded.

**5   Records**
Form QF3-1   Checklist for tender review
Minutes of tender review meeting

Procedure approved by   *A. T. Best*
                        QA Manager
Dated  27/11/98

*Quality Procedure QP3.1*
*Issue 1, 27/11/98*

*ABC Building*
*Construction Co.*

CHECKLIST FOR TENDER REVIEW

Project :

Tender invited by :

|  |  | YES | NO | N/A |
|---|---|---|---|---|
| **1** | **Tender Documents** | | | |
| 1.1 | Are the following documents complete and clear? | | | |
| | Conditions of tender | [ ] | [ ] | [ ] |
| | Specifications | [ ] | [ ] | [ ] |
| | Drawings | [ ] | [ ] | [ ] |
| | Bills of quantities | [ ] | [ ] | [ ] |
| | Site investigation report | [ ] | [ ] | [ ] |
| | Other tender documents | [ ] | [ ] | [ ] |
| 1.2 | Are the following aspects adequately defined? | | | |
| | Construction method/equipment or restrictions thereof | [ ] | [ ] | [ ] |
| | Construction sequence | [ ] | [ ] | [ ] |
| | Special working environment, if required | [ ] | [ ] | [ ] |
| | Special tolerances, if required | [ ] | [ ] | [ ] |
| | Regulatory requirements | [ ] | [ ] | [ ] |
| | Quality assurance requirements | [ ] | [ ] | [ ] |
| **2** | **Site Conditions** | | | |
| 2.1 | Is working environment satisfactory or can be improved? | [ ] | [ ] | [ ] |
| 2.2 | Is access available or can be made? | [ ] | [ ] | [ ] |
| 2.3 | Are required services available or can be arranged? | [ ] | [ ] | [ ] |
| **3** | **Resources** | | | |
| 3.1 | Is manpower available or can be recruited in time? | [ ] | [ ] | [ ] |
| 3.2 | Are plant and equipment available or can be purchased/hired in time? | [ ] | [ ] | [ ] |
| 3.3 | Can subcontracting be arranged? | [ ] | [ ] | [ ] |

*Form QF3-1*
*Issue 1, 27/11/98*

## $\mathcal{A}\,\mathcal{B}\,C\,\mathcal{B}uilding$
## $Construction\ Co.$

Having examined the Company's resources and commitments, I consider that

(a) The Company has the capability to undertake the new project and the tender can be submitted.*

(b) The following matters have to be resolved before the tender can be finalized.*

..................................................................................................

..................................................................................................

..................................................................................................

..................................................................................................

..............................
Contracts Manager

Date

\* *Delete whichever is inapplicable.*

*A B C Building*
*Construction Co.*

### QP 3.2 PROCEDURE FOR CONTRACT REVIEW

**1   Purpose**
    (i)  To ensure that any differences between the contract requirements and those in the tender are resolved before the contract is accepted
    (ii)  To confirm that the Company has the capability to meet the contract requirements

**2   Scope**
Applicable to all construction contracts

**3   Person responsible**
Contracts Manager

**4   Procedure**

4.1   The Contracts Manager and his staff examine the contract documents and compare the requirements with those in the tender.

4.2   Should any differences in requirements be discovered, the Contracts Manager approaches the Architect / Consulting Engineer for resolution.

4.3   Taking into account existing commitments of the Company, the Contracts Manager verifies the Company's capability, both technical and financial, to meet the contractual commitments of the new project.

4.4   The Contracts Manager completes the checklist on Form QF3.2.

4.5   The Contracts Manager arranges a meeting with the General Manager and the Construction Manager to review the contract and to resolve any problem / uncertainty indicated on the checklist. The review meeting is recorded.

**5   Records**
Form QF3-2   Checklist for contract review
Minutes of contract review meeting

Procedure approved by ......... *A. T. Best* .........
                                        QA Manager
Dated  27/11/98

*A B C Building*
*Construction Co.*

## CHECKLIST FOR CONTRACT REVIEW

Project :                                    Client :
Contract No. :            Architect / Engineer :

|   |   | YES | NO | N/A |
|---|---|---|---|---|
| **1** | **Contract Documents** | | | |
| 1.1 | Are contract documents complete and clear? | [ ] | [ ] | [ ] |
| 1.2 | Are contract requirements same as in tender? | [ ] | [ ] | [ ] |
| **2** | **Financial Resources** | | | |
| 2.1 | Is Company funding earmarked? | [ ] | [ ] | [ ] |
| 2.2 | Is necessary bank credit arranged? | [ ] | [ ] | [ ] |
| **3** | **Manpower Resources** | | | |
| 3.1 | Has demand on existing staff been considered? | [ ] | [ ] | [ ] |
| 3.2 | Is necessary training planned? | [ ] | [ ] | [ ] |
| 3.3 | Is necessary recruitment planned? | [ ] | [ ] | [ ] |
| **4** | **Materiel Resources** | | | |
| 4.1 | Are provisions made for major material supplies? | [ ] | [ ] | [ ] |
| 4.2 | Are provisions made for conventional plant req'd? | [ ] | [ ] | [ ] |
| 4.3 | Are provisions made for special plant required? | [ ] | [ ] | [ ] |
| **5** | **Subcontracting** | | | |
| 5.1 | Have subcontracts been negotiated? | [ ] | [ ] | [ ] |
| 5.2 | Have nominated subcontractors been contacted? | [ ] | [ ] | [ ] |

*Form QF3-2*
*Issue 1, 27/11/98*

*A B C Building*
*Construction Co.*

|  |  | YES | NO | N/A |
|---|---|---|---|---|
| **6** | **Others** | | | |
| 6.1 | Is site layout in order? | [ ] | [ ] | [ ] |
| 6.2 | Are transport problems, including site access, resolved? | [ ] | [ ] | [ ] |
| 6.3 | Is design of temporary work arranged? | [ ] | [ ] | [ ] |
| 6.4 | Is preliminary construction programme prepared? | [ ] | [ ] | [ ] |
| 6.5 | Is project quality plan prepared? | [ ] | [ ] | [ ] |

Having examined the Company's resources and commitments, I consider that

(a) The Company has the capability to undertake the new project and the contract can be accepted.*

(b) The following matters have to be resolved before the contract can be accepted.*

................................................................................................

................................................................................................

................................................................................................

................................................................................................

..............................
*Contracts Manager*

Date

\* *Delete whichever is inapplicable.*

*Form QF3-2*
*Issue 1, 27/11/98*

*A B C Building*
*Construction Co.*

## QP 3.3 PROCEDURE FOR VARIATION REVIEW

**1 Purpose**
To ensure that the Company has the capability to meet the revised requirements resulting from a major variation / amendment of the contract

**2 Scope**
Applicable to variation orders involving substantial changes of contractual requirements either in kind or in quantity or both

**3 Person responsible**
Project Manager

**4 Procedure**

4.1 The Project Manager examines the variation order and the revised drawings / specifications, and compares the requirements with those in the original contract documents.

4.2 Should any requirement be inadequately defined or documented, the Project Manager seeks clarification from the Architect / Consulting Engineer.

4.3 The Project Manager verifies that the changes of requirements can be satisfactorily accommodated.

4.4 The Project Manager completes the checklist on Form QF3-3 and forwards it to the Construction Manager for endorsement.

**5 Records**
Form QF3-3   Checklist for variation review

Procedure approved by   *A. T. Best*
                        QA Manager
Dated  27/11/98

*ABC Building*
*Construction Co.*

CHECKLIST FOR VARIATION REVIEW

Project : V. O. No. :

Contract No. :                    V.O. date :

Description of variation :

..................................................................................

Documents / Drawings involved :

..................................................................................

|   |   | YES | NO | N/A |
|---|---|---|---|---|
| **1** | **Revised Drawings/Specifications** | | | |
| 1.1 | Are changes of requirements adequately defined? | [ ] | [ ] | [ ] |
| 1.2 | Are changes of requirements adequately documented? | [ ] | [ ] | [ ] |
| 1.3 | Are changes of requirements compatible with conditions of contract? | [ ] | [ ] | [ ] |
| **2** | **Contractual Implications** | | | |
| 2.1 | Can effects on cost be accommodated? | [ ] | [ ] | [ ] |
| 2.2 | Can effects on time be accommodated? | [ ] | [ ] | [ ] |
| 2.3 | Can subcontracts be varied accordingly? | [ ] | [ ] | [ ] |
| **3** | **Resource Implications** | | | |
| 3.1 | Can material supplies be adjusted? | [ ] | [ ] | [ ] |
| 3.2 | Can plant and equipment use be rescheduled? | [ ] | [ ] | [ ] |
| 3.3 | Can additional manpower demand be satisfied? | [ ] | [ ] | [ ] |

*Form QF3-3*
*Issue 1, 27/11/98*

# A B C Building Construction Co.

Having reviewed the variation order as above, I consider that

(a) The project set-up is capable of accommodating the changes in requirements.*

(b) The following matters have to be resolved before the variation order can be accepted.*

...........................................................................................

...........................................................................................

...........................................................................................

...........................................................................................

..............................
Project Manager

Date

\* *Delete whichever is inapplicable.*

---

Variation order accepted *

Decision deferred pending further investigation by ........................... *

..............................
Construction Manager

Date

\* *Delete whichever is inapplicable.*

Form QF3-3
Issue 1, 27/11/98

*A B C Building*
*Construction Co.*

## QP 5.1 PROCEDURE FOR CONTROL OF DOCUMENTS FOR GENERAL APPLICATION

**1    Purpose**
To control the processes of approval, revision, issuance, distribution and removal of documents which are not project-specific

**2    Scope**
Applicable to the following controlled documents:
- quality system documents including the quality manual, quality procedures, work instructions and forms
- operations manual and related procedures
- safety manual and related instructions
- documents of external origin such as national and international standards, codes of practice, by-laws and statutory regulations

**3    Person responsible**
Quality Assurance Manager / Executive Officer

**4    Procedure**

4.1   With documents of internal origin, the Quality Assurance Manager establishes and maintains a master list indicating the current issue number (and revision number if appropriate) of each document and the date of issue. The list is updated whenever a new issue or a revision of a document is made. The list is placed in the general office in such a way that it is readily available for reference but cannot be taken away or photocopied.

4.2   Prior to issue of a document, the QA Manager reviews and approves the document for adequacy, both in contents and in format. He verifies that the document fulfils the purpose for which it is written. He ensures that the document number, the issue number and issue date are shown on every page, and with a revised document, the amended part is marked with a vertical line drawn along the right-hand margin. He then indicates approval by putting his signature on the document.

4.3   The QA Manager reviews and approves changes to a document in the same manner as above prior to issue.

4.4   When a document contains a number of amendments, say 5, the QA Manager re-issues the whole document incorporating all amendments.

**𝒜 ℬ 𝒞 Building
Construction Co.**

4.5 With documents of external origin, the Executive Officer maintains currency of the documents, replacing an outdated document as soon as the new version is received. The outdated document is either disposed of or stamped with the mark 'SUPERSEDED'. The revision status of each document is checked at least once a year, and arrangement is made immediately to acquire the new version when it is available.

4.6 The Executive Officer is responsible for distribution and retrieval of all controlled documents. For each document, he establishes and maintains a distribution list of controlled copies using Form QF5-1. The controlled copies are numbered in sequence. They are printed on special paper bearing a large mark of 'CONTROLLED COPY' pre-printed in light blue across the page. Use of the special paper is under the strict control of the Executive Officer.

4.7 The Executive Officer distributes the controlled copies of a document, or an amended part of a document, to the designated persons who will place them at convenient locations for their staff to consult. The existing copy, or the part which has been replaced, is promptly removed to assure against unintended use. The obsolete document is returned to the Executive Office for disposal.

4.8 The Executive Officer maintains a stock of current forms in the general office. When a form is amended, the remaining copies of the obsolete form are promptly removed.

4.9 When an obsolete document is returned, the Executive Officer either disposes of it immediately or stamps it with the mark 'SUPERSEDED' if it is retained for future reference.

4.10 An uncontrolled copy of a document supplied for any purpose is stamped with the mark 'UNCONTROLLED COPY - Check currency of issue before use.'

**5 Records**
Form QF5-1 Distribution of controlled copies of document

Procedure approved by ......*𝒜. 𝒯. Best*......
                                   QA Manager
Dated 27/11/98

*ABC Building*
*Construction Co.*

DISTRIBUTION OF CONTROLLED COPIES OF DOCUMENT

Document No. :              Issue No. :

Document title :            Revision No. :

| Copy No. | Document holder | Date distributed | Date returned |
|---|---|---|---|
|  |  |  |  |
|  |  |  |  |
|  |  |  |  |
|  |  |  |  |
|  |  |  |  |
|  |  |  |  |
|  |  |  |  |
|  |  |  |  |
|  |  |  |  |
|  |  |  |  |
|  |  |  |  |
|  |  |  |  |

*Form QF5-1*
*Issue 1, 27/11/98*

*A B C Building*
*Construction Co.*

## QP 5.2 PROCEDURE FOR CONTROL OF DOCUMENTS FOR SPECIFIC PROJECTS

**1   Purpose**
To control the processes of approval, revision, issue, distribution and removal of documents pertaining to a specific project

**2   Scope**
Applicable to the conditions of contract, specifications, contract drawings, bills of quantities and other documents originated from the client, standards and codes of practice released for site use, project quality plan, project-specific procedures, work instructions, working drawings, site investigation reports and miscellaneous data sheets

**3   Person responsible**
Contracts Manager / Project Manager / Quality Assurance Officer

**4   Procedure**

4.1   The Contracts Manager is responsible for control of documents and drawings supplied by the client. For each project, he establishes and maintains a distribution list of contract documents and drawings using Form QF5-2.

4.2   When a revised document / drawing is received from the client, the Contracts Manager records it on the list before distribution and ensures prompt return of the document / drawing which is made obsolete. The obsolete document / drawing is stamped as 'SUPERSEDED' before storage.

4.3   With documents and drawings prepared internally for the project, such as the project quality plan, technical procedures, work instructions for special tasks and working drawings, the Project Manager (or the Engineer as appropriate) reviews and approves such document / drawing before release for use. An approved document / drawing carries the signature of the authorized person. Changes to a document or drawing are reviewed and approved by the same person or the current incumbent of the same position.

*Quality Procedure QP5.2*
*Issue 1, 27/11/98*

*A B C Building*
*Construction Co.*

4.4 On site, the Quality Assurance Officer is responsible for document control. He establishes and maintains a register of documents and drawings kept in the site office using Form QF5-3. Controlled copies of such are issued to those persons (including subcontractors) who need them for their operation. Distribution and retrieval of controlled copies are recorded on Form QF5-2. The disposition of the superseded copies is at the discretion of the Project Manager.

4.5 An uncontrolled copy of a document / drawing supplied for any purpose is stamped with the mark 'UNCONTROLLED COPY - Check currency of issue before use.'

5 **Records**
Form QF5-2 Distribution of project documents and drawings
Form QF5-3 Register of project documents and drawings

Procedure approved by ......*A. T. Best*......
*QA Manager*
Dated 27/11/98

*A B C Building*
*Construction Co.*

## DISTRIBUTION OF PROJECT DOCUMENTS / DRAWINGS

Project :   Client :
Contract No. :   Architect / Engineer :

| Document / Drawing title | Issue No. | Revision No. | Distributed to | Date distributed | Date returned |
|---|---|---|---|---|---|
| | | | | | |
| | | | | | |
| | | | | | |
| | | | | | |
| | | | | | |
| | | | | | |
| | | | | | |
| | | | | | |
| | | | | | |
| | | | | | |
| | | | | | |
| | | | | | |

*Form QF5-2*
*Issue 1, 27/11/98*

*ABC Building*
*Construction Co.*

REGISTER OF PROJECT DOCUMENTS AND DRAWINGS

Project :                            Client :

Contract No. :              Architect / Engineer :

| Document title / Drawing title | Issue No. | Issue date | Revision date | | | |
|---|---|---|---|---|---|---|
| | | | 1 | 2 | 3 | 4 |
| | | | | | | |
| | | | | | | |
| | | | | | | |
| | | | | | | |
| | | | | | | |
| | | | | | | |
| | | | | | | |
| | | | | | | |
| | | | | | | |
| | | | | | | |
| | | | | | | |
| | | | | | | |

*Form QF5-3*
*Issue 1, 27/11/98*

*A B C Building*
*Construction Co.*

**QP 6.1   PROCEDURE FOR EVALUATION OF SUBCONTRACTORS**

**1   Purpose**
   (i)   To maintain a list of acceptable subcontractors for a trade or service
   (ii)  To maintain quality records of acceptable subcontractors

**2   Scope**
Applicable to subcontracting of all kinds

**3   Person responsible**
Construction Manager

**4   Procedure**

4.1   The Construction Manager maintains a list of acceptable subcontractors for each trade or service, using Form QF6-1. The list is subject to document control as per Quality Procedure QP 5.1 Controlled copies of the list are distributed to the Quality Assurance Manager, Contracts Manager, Purchasing Officer and Executive Officer.

4.2   The Construction Manager evaluates any potential subcontractor before inclusion on the list. The evaluation is based on the subcontractor's resources, past performance, quality management, and references if appropriate. The result of evaluation is recorded on Form QF6-2.

4.3   The Project Manager reviews the performance of each subcontractor annually and at completion of the subcontract, and submits the results on Form QF6-2 to the Contracts Manager. The Contracts Manager adds his comments to the form which then goes to the Construction Manager.

4.4   Based on the reports from all projects in which the particular subcontractor is involved, the Construction Manager either retains the subcontractor on the list or removes the subcontractor from the list.

**5   Records**
Form QF6-1   List of acceptable subcontractors for ...............
Form QF6-2   Evaluation of subcontractor

Procedure approved by  *A. T. Best*
.........................
QA Manager

Dated  27/11/98

*Quality Procedure QP6.1*
*Issue 1, 27/11/98*

# A B C Building Construction Co.

## LIST OF ACCEPTABLE SUBCONTRACTORS FOR ..................
*trade / service*

| Company name and address | Contact person | $ limit of transaction | Last job (month/year) | Date evaluated | Date to be re-evaluated | Remarks |
|---|---|---|---|---|---|---|
| | | | | | | |
| | | | | | | |
| | | | | | | |
| | | | | | | |
| | | | | | | |
| | | | | | | |
| | | | | | | |

Note : *After re-evaluation, cross out existing entry and re-enter at bottom of list.*

*Form QF6-1*
*Issue 1, 27/11/98*

*A B C Building*
*Construction Co.*

## EVALUATION OF SUBCONTRACTOR

Subcontractor :            Subcontract :
Trade / service :          Period covered :

|    |                                              | GOOD | AVERAGE | POOR | N/A |
|----|----------------------------------------------|------|---------|------|-----|
| 1  | Human resources                              | [ ]  | [ ]     | [ ]  | [ ] |
| 2  | Plant and equipment                          | [ ]  | [ ]     | [ ]  | [ ] |
| 3  | Workmanship                                  | [ ]  | [ ]     | [ ]  | [ ] |
| 4  | On-time delivery                             | [ ]  | [ ]     | [ ]  | [ ] |
| 5  | Care of Company material                     | [ ]  | [ ]     | [ ]  | [ ] |
| 6  | Care of Company plant                        | [ ]  | [ ]     | [ ]  | [ ] |
| 7  | Occupational health                          | [ ]  | [ ]     | [ ]  | [ ] |
| 8  | Safety                                       | [ ]  | [ ]     | [ ]  | [ ] |
| 9  | Tidiness                                     | [ ]  | [ ]     | [ ]  | [ ] |
| 10 | Quality management                           | [ ]  | [ ]     | [ ]  | [ ] |
| 11 | Financial standing                           | [ ]  | [ ]     | [ ]  | [ ] |
| 12 | References (for new subcontractor only)      | [ ]  | [ ]     | [ ]  | [ ] |
| 13 | Other comments                               |      |         |      |     |

..................................................................................

..................................................................................

..................................................................................

............................
*Project Manager*

Date

*Form QF6-2*
*Issue 1, 27/11/98*

## ABC Building Construction Co.

Comments on contractual obligations :

..................................................................................................

..................................................................................................

..................................................................................................

..............................
Contracts Manager

Date

---

Decision :

(a) Performance satisfactory; allowed to continue work *

(b) Performance unsatisfactory; allowed to complete current subcontract *

(c) Performance unsatisfactory; subcontract to be terminated *

(d) To be retained on list of acceptable subcontractors *

(e) To be removed from list of acceptable subcontractors *

[ For new subcontractor ]

(f) To be included on list of acceptable subcontractors with transaction limit $ ............... *

(g) Not to be included on list of acceptable subcontractors *

..............................
Construction Manager

Date

\* *Delete whichever is inapplicable.*

Form QF6-2
Issue 1, 27/11/98

*A B C Building*
*Construction Co.*

## QP 6.2 PROCEDURE FOR EVALUATION OF SUPPLIERS

**1 Purpose**
To maintain a list of acceptable suppliers of material, appliance or equipment

**2 Scope**
Applicable to purchase of major items of material, appliance and equipment

**3 Person responsible**
Construction Manager

**4 Procedure**

4.1 The Construction Manager maintains a list of acceptable suppliers for each major item of material, appliance or equipment, using Form QF6-3. The list is subject to document control as per Quality Procedure QP 5.1. Controlled copies of the list are distributed to the Quality Assurance Manager, Contracts Manager, Purchasing Officer and Executive Officer.

4.2 The Construction Manager evaluates any potential supplier before inclusion on the list. The evaluation is based on the supplier's range and quality of supply, past performance, quality management, and references if appropriate. The result of evaluation is recorded on Form QF6-4.

4.3 The Project Manager reviews the performance of each supplier annually and at completion of the supply period, and submits the results on Form QF6-4 to the Contracts Manager. The Contracts Manager adds his comments to the form which goes to the Construction Manager.

4.4 Based on the reports from all projects in which the particular supplier is involved, the Construction Manager either retains the supplier on the list or removes the supplier from the list.

**5 Records**
Form QF6-3   List of acceptable suppliers of ............
Form QF6-4   Evaluation of supplier

Procedure approved by   *A. T. Best*
............................
*QA Manager*

Dated 27/11/98

*ABC Building Construction Co.*

LIST OF ACCEPTABLE SUPPLIERS OF ..................
*material / appliance*

| Company name and address | Contact person | $ limit of transaction | Last job (month/year) | Date evaluated | Date to be re-evaluated | Remarks |
|---|---|---|---|---|---|---|
|  |  |  |  |  |  |  |
|  |  |  |  |  |  |  |
|  |  |  |  |  |  |  |
|  |  |  |  |  |  |  |
|  |  |  |  |  |  |  |
|  |  |  |  |  |  |  |
|  |  |  |  |  |  |  |

Note : *After re-evaluation, cross out existing entry and re-enter at bottom of list.*

Form QF6-3
Issue 1, 27/11/98

## $\mathcal{ABC}$ $\mathcal{B}uilding$ Construction Co.

### EVALUATION OF SUPPLIER

Supplier :  Purchase order :
Item supplied :  Period covered :

|   |   | GOOD | AVERAGE | POOR | N/A |
|---|---|---|---|---|---|
| 1 | Quality of supply | [ ] | [ ] | [ ] | [ ] |
| 2 | Range of supply | [ ] | [ ] | [ ] | [ ] |
| 3 | On-time delivery | [ ] | [ ] | [ ] | [ ] |
| 4 | Technical backup | [ ] | [ ] | [ ] | [ ] |
| 5 | After-sale service | [ ] | [ ] | [ ] | [ ] |
| 6 | Quality management | [ ] | [ ] | [ ] | [ ] |
| 7 | Financial standing | [ ] | [ ] | [ ] | [ ] |
| 8 | References (for new supplier only) | [ ] | [ ] | [ ] | [ ] |

7  Other comments

..................................................................................

..................................................................................

..................................................................................

............................
*Project Manager*

Date

*Form QF6-4*
*Issue 1, 27/11/98*

## ABC Building Construction Co.

Comments on contractual obligations :

..................................................................................................

..................................................................................................

..................................................................................................

..............................
Contracts Manager

Date

---

Decision :

(a) Supply satisfactory; allowed to continue supply *

(b) Supply unsatisfactory; allowed to complete current purchase order *

(c) Supply unsatisfactory; purchase order to be terminated *

(d) To be retained on list of acceptable suppliers *

(e) To be removed from list of acceptable suppliers *

[ For new supplier ]

(f) To be included on list of acceptable suppliers

with transaction limit $ ............... *

(g) Not to be included on list of acceptable suppliers *

..............................
Construction Manager

Date

\*    *Delete whichever is inapplicable.*

Form QF6-4
Issue 1, 27/11/98

*A B C Building*
*Construction Co.*

**QP 7**  PROCEDURE FOR CONTROL OF CLIENT-SUPPLIED ITEMS

**1**  **Purpose**
To control verification, storage and use of items supplied by client

**2**  **Scope**
Applicable to materials, components, sub-assemblies and appliances supplied by client for incorporation into the permanent works

**3**  **Person responsible**
General Foreman

**4**  **Procedure**
4.1  The General Foreman receives delivery of the item and verifies that it is of the type, model and quantity specified in the contract.

4.2  The General Foreman or his delegate inspects the item as soon as possible after receipt. The receiving inspection follows Quality Procedure QP10.1 and includes filling in of Form QF10-1. Any damage or defect found in the material or malfunctioning of the appliance is reported to the Project Manager who will then notify the client accordingly.

4.3  The General Foreman or his delegate handles and stores the item in accordance with Quality Procedure QP15 and/or the client's instructions. Any item that is lost, damaged or deteriorated during handling or storage is recorded and reported to the client.

**5**  **Records**
Form QF10-1   Material receipt and issue

Procedure approved by  *A. T. Best*
.........................
*QA Manager*
Dated  27/11/98

## $\mathcal{ABC}$ $\mathcal{B}uilding$ Construction Co.

### MATERIAL RECEIPT AND ISSUE

Project :                           Item received :

Contract No. :                      Supplier :

---

### INWARD

Quantity :

Date received :

Identification :
*(if applicable)*

Documents : Not required    [  ]
               Complete          [  ]
               Incomplete        [  ]

Type/Grade : As specified   [  ]
              Different           [  ]

Condition : Satisfactory    [  ]
            Unsatisfactory   [  ]

Test :   Not required    [  ]
        Pending            [  ]
        Satisfactory       [  ]
        Unsatisfactory   [  ]

This item can be issued for use.*

This item may be issued for use subject to replacement.*
*(Project Manager to initial.)*

This item is withheld pending test results.*

This item must not be issued for use.*

..............................    ..........
*General Foreman*         Date

\* Delete whichever is inapplicable.

### OUTWARD

Quantity issued :

Where used, if known :

Received by :
..............................
Signature & date

Quantity issued :

Where used, if known :

Received by :
..............................
Signature & date

Quantity issued :

Where used, if known :

Received by :
..............................
Signature & date

Total issued   =

Wastage       =   _____

Overall total  =

This item has been completely issued.

..............................    ..........
*Storekeeper*            Date

..............................    ..........
*General Foreman*         Date

---

Form QF10-1                                    Page 1 of 1
Issue 1, 27/11/98

*A B C Building*
*Construction Co.*

**QP 8    PROCEDURE FOR PRODUCT IDENTIFICATION AND TRACEABILITY**

**1    Purpose**
To identify finished or semi-finished product from production through to installation

**2    Scope**
Applicable to precast or prefabricated units and sub-assemblies

**3    Person responsible**
Foreman

**4    Procedure**

4.1    The Foreman supervising the production puts an indelible mark (e.g. beam mark or lot number) on the precast or prefabricated units, relating the unit to the corresponding contract drawing and/or indicating the location where the units will be installed.

4.2    Where traceability of the units is specified in the contract, the Foreman puts a unique mark on each individual unit and keeps a record of the marks together with details of production of the corresponding units, such as date of production and inspection/testing status.

**5    Records**
List of marks of individual units with production details

Procedure approved by ......*A. T. Best*......
                           QA Manager
Dated  27/11/98

# A B C Building Construction Co.

## QP 9  PROCEDURE FOR PROCESS CONTROL

**1  Purpose**
 (i)  To identify and plan the site activities ahead of time
 (ii) To ensure that these activities are performed under controlled conditions

**2  Scope**
Applicable to construction and erection processes, including erection of temporary works; also applicable to production of precast or prefabricated units

**3  Person responsible**
Project Manager

**4  Procedure**

4.1  Based on the construction programme included in the project quality plan, the Project Manager prepares a bi-weekly programme of construction activities, using Form QF9-1.

4.2  The Project Manager liaises with the head-office to arrange for the supply of materials, labour and equipment on site at the appropriate times. He also coordinates the subcontractors to work to the programme.

4.3  The Project Manager, or the Engineer working under his direction, prepares in advance work instructions for the processes identified in the project quality plan as requiring such documents. The work instructions should describe the method and sequence of work, equipment to be used, suitable working environment and criteria of workmanship. The quality of work to be achieved is based on the specifications of the contract and the relevant national standard or code of practice. If the work is subcontracted, the work instruction is issued to the subcontractor unless the subcontractor has his own work instruction which is considered suitable for the purpose.

*A B C Building*
*Construction Co.*

4.4 The General Foreman, under the direction of the Project Manager, organizes the resources necessary for the scheduled activities on a daily basis, ensuring that the labour carrying out a particular process has adequate skill and experience. For special processes, only labour with appropriate qualification and/or training is assigned to carry out the job. (For example, only licensed welders are assigned to perform welding.) This applies to direct labour as well as subcontracted labour.

4.5 The General Foreman, together with the Engineer, exercises overall supervision of the activities on site. He assigns the foremen to supervise individual processes so as to ensure that the work instructions or normal practices of the Company are followed.

4.6 The General Foreman enters the daily progress of work against the scheduled activities on the bi-weekly programme. He reports any significant delay, and suggests any necessary rescheduling to the Project Manager.

4.7 The Quality Assurance Officer prepares a schedule of equipment maintenance, arranges for maintenance as scheduled and keeps record of such.

4.8 The Project Manager holds a project meeting at appropriate intervals (normally every two weeks) to review the progress of the project and any difficulties encountered. The meeting is attended by the General Foreman, the Quality Assurance Officer, the Engineer (if required), the subcontractors involved and other persons with special duties. The project meeting is minuted.

4.9 The Project Manager updates the project quality plan whenever necessary and forwards it to the Quality Assurance Manager for record and controlled distribution.

**5  Records**
Form QF9-1   Bi-weekly programme
Minutes of project meetings
Equipment maintenance schedule

Procedure approved by   *A. T. Best*
.........................
QA Manager

Dated  27/11/98

*ABC Building
Construction Co.*

### BI - WEEKLY PROGRAMME
Period starting .......................

Project :                                    Client :

Contract No. :                        Architect / Engineer :

| Operation | Day / Date | M | T | W | T | F | S | S | M | T | W | T | F | S | S |
|---|---|---|---|---|---|---|---|---|---|---|---|---|---|---|---|
| | | | | | | | | | | | | | | | | |
| | | | | | | | | | | | | | | | | |
| | | | | | | | | | | | | | | | | |
| | | | | | | | | | | | | | | | | |
| | | | | | | | | | | | | | | | | |
| | | | | | | | | | | | | | | | | |
| | | | | | | | | | | | | | | | | |
| | | | | | | | | | | | | | | | | |
| | | | | | | | | | | | | | | | | |
| | | | | | | | | | | | | | | | | |
| | | | | | | | | | | | | | | | | |

*Form QF9-1*
*Issue 1, 27/11/98*

# ABC Building Construction Co.

## QP 10.1 PROCEDURE FOR RECEIVING INSPECTION AND TESTING

**1 Purpose**
  (i) To ensure that items received on site are inspected and/or tested before being released for use
  (ii) To specify actions in case of release for urgent use prior to verification
  (iii) To keep a record of material receipt and issue on site

**2 Scope**
Applicable to incoming materials (which include components and appliances) and equipment, including client supplied items but excluding minor items in small quantities

**3 Person responsible**
General Foreman

**4 Procedure**

4.1 On receipt of an item of material or equipment, the General Foreman or his delegate checks its identification and quantity against the delivery docket and the purchase order. He also checks the accompanying documents such as mill certificate, test report and installation manual.

4.2 The General Foreman or his delegate inspects the item as soon as possible after receipt and records the type, model, quantity and condition of the item on Form QF10-1. Where so indicated in the project quality plan, he arranges for the item to be sampled by an authorized person and tested by an independent testing laboratory. The item is withheld pending test results.

4.3 When the item has passed the inspection and/or testing, the General Foreman authorizes its release for use or installation by signing off the 'inward' portion of Form QF10-1.

4.4 If the item does not pass the inspection and/or testing, it is dealt with as nonconforming supply in accordance with Quality Procedure QP13.1.

## A B C Building Construction Co.

4.5 When instructed by the General Foreman, the Storekeeper issues the item, or part of it, and records the quantity issued on the 'outward' portion of Form QF10-1. After the item is completely issued, he balances the inward quantity with the outward quantity and signs off the 'outward' portion of Form QF10-1 which is then endorsed by the General Foreman. Any extraordinary wastage is reported to the Project Manager.

4.6 Should the item be urgently needed for use or installation before inspection and/or testing is completed, the General Foreman seeks the approval of the Project Manager who will, at his discretion, sign for the release of the item. The exact location where the item is used or installed must be recorded on Form QF10-1.

4.7 If traceability of the item is specified in the contract, record is maintained on Form QF10-1 of the unique mark of each individual unit and the exact location where it is used or installed.

**5 Records**
Form QF10-1 Material receipt and issue

Procedure approved by *A. T. Best*
.........................
*QA Manager*

Dated 27/11/98

## $\mathcal{A} \mathcal{B} C$ Building Construction Co.

### MATERIAL RECEIPT AND ISSUE

Project :  Item received :

Contract No. :  Supplier :

---

### INWARD

Quantity :

Date received :

Identification :
*(if applicable)*

Documents : Not required [ ]
             Complete [ ]
             Incomplete [ ]

Type/Grade : As specified [ ]
             Different [ ]

Condition : Satisfactory [ ]
           Unsatisfactory [ ]

Test : Not required [ ]
      Pending [ ]
      Satisfactory [ ]
      Unsatisfactory [ ]

This item can be issued for use.*

This item may be issued for use subject to replacement.*
*(Project Manager to initial.)*

This item is withheld pending test results.*

This item must not be issued for use.*

..............................  ............
General Foreman     Date

\* *Delete whichever is inapplicable.*

### OUTWARD

Quantity issued :

Where used, if known :

Received by :
..............................
Signature & date

Quantity issued :

Where used, if known :

Received by :
..............................
Signature & date

Quantity issued :

Where used, if known :

Received by :
..............................
Signature & date

Total issued =

Wastage =
           _____
Overall total =

This item has been completely issued.

..............................  ............
Storekeeper     Date

..............................  ............
General Foreman     Date

---

Form QF10-1
Issue 1, 27/11/98

# ABC Building Construction Co.

## QP 10.2 PROCEDURE FOR IN-PROCESS INSPECTION AND TESTING

**1  Purpose**
To prevent unsatisfactory work being built upon or covered up

**2  Scope**
Applicable to construction work by direct labour or subcontractors, including precast and prefabricated work

**3  Person responsible**
Engineer - for engineering work and special processes
General Foreman - for general building work

**4  Procedure**

4.1  The Engineer / General Foreman arranges for inspection and/or testing of construction work (either by direct labour or by a subcontractor) at the various inspection / test points indicated in the inspection and test plan. Prior notice of the arrangement is given to the designated inspector. The designated inspector may be the Engineer / General Foreman himself, another member of supervisory staff of the Company or an authorized person from the Architect's / Consulting Engineer's office, as indicated in the inspection and test plan.

4.2  If the inspector is satisfied with the quality of work, he indicates his approval by signing at the corresponding point of the inspection and test plan.

4.3  If a minor defect is noted, which can be rectified within 24 hours and is not a recurrence, the Engineer / General Foreman instructs the workers or the subcontractor to carry out the rectification. Any work that is rectified is re-inspected and/or re-tested. If a major defect is noted, the work is dealt with as nonconforming work in accordance with Quality Procedure QP13.2.

4.4  For such work not covered by an inspection and test plan, the Engineer / General Foreman carries out in-process inspection in the usual manner and, at his discretion, allows the construction process to proceed or otherwise.

## *A B C Building*
## *Construction Co.*

4.5  The Engineer / General Foreman records the in-process inspection and/or testing on Form QF10-2.

**5  Records**
Form QF10-2   Record of inspection

Procedure approved by  *A. T. Best*
.........................
QA Manager

Dated  27/11/98

## $\mathcal{ABC}$ $\mathcal{B}uilding$ $Construction$ $Co.$

### RECORD OF INSPECTION

Project :                                   Client :

Contract No. :              Architect / Engineer :

| Date of inspection | Location of inspection | Trade / Subcontract | Pass or Fail | Comments ( if fail ) | Inspector's signature |
|---|---|---|---|---|---|
|  |  |  |  |  |  |
|  |  |  |  |  |  |
|  |  |  |  |  |  |
|  |  |  |  |  |  |
|  |  |  |  |  |  |
|  |  |  |  |  |  |
|  |  |  |  |  |  |
|  |  |  |  |  |  |
|  |  |  |  |  |  |
|  |  |  |  |  |  |
|  |  |  |  |  |  |
|  |  |  |  |  |  |
|  |  |  |  |  |  |
|  |  |  |  |  |  |

*A B C Building
Construction Co.*

## QP 10.3 PROCEDURE FOR FINAL INSPECTION AND TESTING

**1 Purpose**
To complete the evidence of conformance of the finished work to the specified requirements

**2 Scope**
Applicable to construction work by direct labour and subcontractors, including precast and prefabricated work

**3 Person responsible**
Project Manager

**4 Procedure**

4.1 The Project Manager or his delegate arranges for final inspection and/or testing of the construction work that has been completed by direct labour or by a subcontractor. Prior notice of the arrangement is given to the designated inspector. The designated inspector may be the Project Manager or his delegate, another member of supervisory staff of the Company or an authorized person from the Architect's / Consulting Engineer's office, as indicated in the inspection and test plan.

4.2 The inspector also checks the inspection and test plan for signatures at the various inspection / test points to ensure that all in-process inspection and/or tests have been carried out and the results meet specified requirements.

4.3 If the inspector is satisfied with the quality of work, he indicates his approval by signing off the inspection and test plan. The work is not considered as duly completed until this is accomplished.

4.4 If a minor defect is noted, the Project Manager takes appropriate action to put it right or instructs the subcontractor to do so. Any work that is rectified is re-inspected and/or re-tested. If a major defect is noted, the work is dealt with as nonconforming work in accordance with Quality Procedure QP13.2.

4.5 For such work not covered by an inspection and test plan, the Project Manager or his delegate carries out final inspection and/or testing in the usual manner and, at his discretion, accepts or rejects the work.

*A B C Building*
*Construction Co.*

4.6 The Project Manager or his delegate records the final inspection and/or testing on Form QF10-2.

**5 Records**
Form QF10-2   Record of inspection

*A. T. Best*
Procedure approved by ........................
*QA Manager*
Dated 27/11/98

*A B C Building*
*Construction Co.*

## RECORD OF INSPECTION

Project :                                    Client :

Contract No. :                    Architect / Engineer :

| Date of inspection | Location of inspection | Trade / Subcontract | Pass or Fail | Comments ( if fail ) | Inspector's signature |
|---|---|---|---|---|---|
|  |  |  |  |  |  |
|  |  |  |  |  |  |
|  |  |  |  |  |  |
|  |  |  |  |  |  |
|  |  |  |  |  |  |
|  |  |  |  |  |  |
|  |  |  |  |  |  |
|  |  |  |  |  |  |
|  |  |  |  |  |  |
|  |  |  |  |  |  |
|  |  |  |  |  |  |
|  |  |  |  |  |  |

*Form QF10-2*
*Issue 1, 27/11/98*

*A B C Building
Construction Co.*

## QP 11 PROCEDURE FOR CONTROL OF MEASURING AND TEST EQUIPMENT

**1 Purpose**
To control, calibrate and maintain measuring and test equipment

**2 Scope**
Applicable to surveying instruments, weighbridges, prestressing jacks, torque wrenches, testing devices, weighing scales and measuring tapes, both self-owned and hired

**3 Person responsible**
Quality Assurance Officer

**4 Procedure**

4.1 The Quality Assurance Officer establishes and maintains a Record of Equipment Calibration on Form QF11-1 covering the measuring and test equipment on site.

4.2 On receipt of a piece of equipment delivered to the site, either from the manufacturer or from another site, the QA Officer verifies, by referring to the project quality plan if appropriate, that it is capable of the necessary accuracy and precision. He also checks its calibration status (except for measuring tapes) and working condition. He then makes an entry of the piece of equipment into the Record of Equipment Calibration.

4.3 The QA Officer arranges for calibration (and adjustment as appropriate) of each piece of equipment (except measuring tapes), by an accredited organization if available, before it is first used unless it comes with a certificate of calibration, and then at an interval as recommended by the manufacturer or indicated in the project quality plan, and whenever there is suspicion that the equipment is out of calibration.

4.4 The QA Officer places a sticker or tag on each piece of equipment indicating the serial number, the ranges calibrated where appropriate, the date of calibration and the date of next calibration.

**ABC Building
Construction Co.**

4.5 When a piece of equipment is found to be out of calibration, the QA Officer immediately removes the equipment from service and places a warning label or tag on it to avoid inadvertent use. He then assesses the validity of previous results obtained with the equipment and reports the outcome to the Project Manager. The incident and subsequent action are entered as remarks in the Record of Equipment Calibration.

4.6 The QA Officer checks the measuring tapes for wear and tear every month. Should a tape show significant wear and tear, he immediately removes it from service, otherwise he puts on it a colour sticker representing the current month.

4.7 The QA Officer keeps custody of all portable equipment, and records the release and return of the equipment in a log-book.

4.8 With equipment belonging to a subcontractor (except measuring tapes), the QA Officer conducts a suitability check as in 4.2 before it is used. He also ensures that the equipment carries an indicator showing the calibration status. Unless the subcontractor keeps a record of its calibration, the QA Officer includes the equipment in his Record of Equipment Calibration and indicates in the remarks column to whom it belongs. He reminds the subcontractor to arrange for calibration whenever it is due.

## 5 Records

Form QF11-1: Record of Equipment Calibration
Reports of equipment calibration

Procedure approved by ......*A. T. Best*......
                              QA Manager

Dated 27/11/98

## ABC Building Construction Co.

### RECORD OF EQUIPMENT CALIBRATION

Project :  Client :
Contract No. :  Architect / Engineer :

| Equipment type | Serial No. | Calibration interval | Date of calibration 1 | 2 | 3 | Remarks |
|---|---|---|---|---|---|---|
| | | | | | | |
| | | | | | | |
| | | | | | | |
| | | | | | | |
| | | | | | | |
| | | | | | | |
| | | | | | | |
| | | | | | | |
| | | | | | | |
| | | | | | | |

Form QF11-1
Issue 1, 27/11/98

*A B C Building*
*Construction Co.*

**QP 12 PROCEDURE FOR IDENTIFICATION OF INSPECTION AND TEST STATUS**

**1  Purpose**
To identify separately conforming products and nonconforming products

**2  Scope**
Applicable to precast concrete units and other prefabricated components such as steel sub-assemblies

**3  Person responsible**
General Foreman

**4  Procedure**

4.1  The General Foreman keeps custody of copies of the following label.

> **INSPECTION PASSED**
>
> ANOTHER QUALITY PRODUCT BY
> ABC BUILDING CONSTRUCTION CO.

4.2  The General Foreman issues the labels to the foreman supervising the production who will stick them on the items that have passed the final inspection and testing.

4.3  The Foreman puts an indelible mark (e.g. beam mark or lot number) on the precast or prefabricated units, relating the unit to the corresponding contract drawing and/or indicating the location where the units will be installed. (Refer to Quality Procedure QP8.)

4.4  Should any item fail the final inspection and testing, the foreman segregates it from the rest. Disposition of the nonconforming product will be in accordance with Quality Procedure QP13.1.

4.5  The foreman ensures that only those items with the label are dispatched.

**5  Records**
Nil

Procedure approved by  *A. T. Best*
                        QA Manager
Dated 27/11/98

Quality Procedure QP12
Issue 1, 27/11/98

*ABC Building
Construction Co.*

### QP 13.1 PROCEDURE FOR CONTROL OF NONCONFORMING SUPPLY

**1 Purpose**
(i) To ensure that material supply that does not conform to specified requirements is prevented from unintended use or installation
(ii) To define responsibility for review and authority for the disposition of nonconforming supply

**2 Scope**
Applicable to purchased material, components and appliances which do not pass the receiving inspection and testing, except minor items in small quantities

**3 Person responsible**
Project Manager

**4 Procedure**

4.1 The General Foreman identifies the nonconforming item by physical means and segregates it from the others.

4.2 The General Foreman refers the nonconforming item to the Project Manager for disposition.

4.3 The Project Manager reviews the extent and severity of the nonconformity and decides on the disposition of the nonconforming item. The disposition may be concessional acceptance, regrading for alternative use, repair, reject or scrap.

4.4 The Project Manager issues a Notice of Nonconforming Supply (Form QF13-1) to the supplier, copied to the Purchasing Officer.

4.5 The Purchasing Officer records the incident of nonconformance in the List of Nonconforming Supply (Form QF13-2).

4.6 The Purchasing Officer initiates corrective action in accordance with Quality Procedure QP14.1 if he considers it necessary.

4.7 Minor items in small quantities that are nonconforming are exempted from this procedure. In such cases, the Project Manager or the General Foreman acting on his behalf decides on the disposition of the nonconforming item and informs the Purchasing Officer accordingly.

*A B C Building*
*Construction Co.*

5   **Records**
    Form QF13-1   Notice of nonconforming supply
    Form QF13-2   List of nonconforming supply

                         *A. T. Best*
Procedure approved by ..........................
                         *QA Manager*

Dated   27/11/98

*A B C Building*
*Construction Co.*

## NOTICE OF NONCONFORMING SUPPLY

To :                                    Purchase order :

Attention :                          Item(s) supplied :

Ref. No. :

---

The supply described below is not in compliance with the purchase order and will be subject to disposition as indicated.

Consignment identification :

Date of delivery :

Where delivered :

Description of nonconformity :

Contract term / specification not satisfied :

Disposition of nonconforming item(s) :

Deadline for action :

You are hereby instructed to carry out the indicated action by the due date. Proposal for alternative action will be considered if it is submitted not later than ..........................

..........................
*Project Manager*

Date

c.c.: Purchasing Officer

*Form QF13-1*
*Issue 1, 27/11/98*

*A B C Building
Construction Co.*

## LIST OF NONCONFORMING SUPPLY

| Ref. No. | Date received | Item and quantity | Supplier | Supplied to | Non-conformity | Disposition |
|---|---|---|---|---|---|---|
| | | | | | | |
| | | | | | | |
| | | | | | | |
| | | | | | | |
| | | | | | | |
| | | | | | | |
| | | | | | | |
| | | | | | | |
| | | | | | | |
| | | | | | | |
| | | | | | | |
| | | | | | | |
| | | | | | | |

*Form QF13-2
Issue 1, 27/11/98*

*ABC Building*
*Construction Co.*

**QP 13.2  PROCEDURE FOR CONTROL OF NONCONFORMING WORK**

**1   Purpose**
  (i)  To ensure that finished or semi-finished work which does not conform to specified requirements is prevented from being unintentionally covered or built upon
  (ii) To define responsibility for review and authority for the disposition of nonconforming work

**2   Scope**
Applicable to construction work which does not pass the in-process or final inspection and testing

**3   Person responsible**
Project Manager

**4   Procedure**

4.1  The General Foreman / Engineer identifies the nonconforming work on the working drawing with indelible ink. If it is practical, the nonconforming work on location is also marked with paint.

4.2  The Project Manager, assisted by the Engineer and the Quality Assurance Officer, reviews the extent and severity of the nonconformity.

4.3  Based on the outcome of the review, the Project Manager works out a proposal for the disposition of the nonconforming work. The disposition may be concessional acceptance, repair or rework.

4.4  The Project Manager discusses the proposed disposition with the Architect / Consulting Engineer and implements a mutually agreed solution.

4.5  If the work is subcontracted, the Project Manager issues a Notice of Nonconforming Work to the subcontractor, using Form QF13-3.

4.6  The Project Manager ensures that any repaired work is re-inspected and/or re-tested in accordance with the inspection and test plan.

4.7  The Project Manager records the nonconformance in the Record of Nonconforming Work (Form QF13-4) and signs off the entry after its disposition.

Quality Procedure QP13.2
Issue 1, 27/11/98

*A B C Building*
*Construction Co.*

4.8 The Project Manager takes corrective action in accordance with Quality Procedure QP14.1 if he considers it necessary.

4.9 The Project Manager prepares a Report of Nonconforming Work on Form QF13-5 and submits it to the Quality Assurance Manager.

4.10 Minor defects, which can be rectified within 24 hours and are not a recurrence, are exempted from this procedure. In such cases, the General Foreman / Engineer conducting the inspection or testing instructs the workers or the subcontractor to carry out the rectification. Any work that is rectified is re-inspected and/or re-tested.

**5 Records**
   Form QF13-3  Notice of nonconforming work
   Form QF13-4  Record of nonconforming work
   Form QF13-5  Report of nonconforming work

Procedure approved by  *A. T. Best*
                        QA Manager
Dated 27/11/98

*ABC Building*
*Construction Co.*

## NOTICE OF NONCONFORMING WORK

To :                                    Subcontract :

Attention :                             Site :

Ref. No. :

_____

The activity / work described below is not in compliance with the subcontract and remedial measures are required.

Location of nonconformity:

Description of nonconformity :

Contract term / specification not satisfied :

Repair/rework required :

Deadline for completion :

You are hereby instructed to carry out the indicated action by the due date. Proposal for alternative action, including a time schedule for implementation, will be considered if it is submitted not later than   ......................


..............................
*Project Manager*

Date

*Form QF13-3*
*Issue 1, 27/11/98*

## $\mathcal{ABC}$ $\mathcal{B}uilding$ $Construction$ $Co.$

### RECORD OF NONCONFORMING WORK

Project :  Client :
Contract No. :  Architect / Engineer :

| Ref. No. | Date of occurrence | Location of occurrence | Nature of nonconformity | Disposition | Signature and date |
|---|---|---|---|---|---|
| | | | | | |
| | | | | | |
| | | | | | |
| | | | | | |
| | | | | | |
| | | | | | |
| | | | | | |
| | | | | | |
| | | | | | |
| | | | | | |
| | | | | | |
| | | | | | |

*Form QF13-4*
*Issue 1, 27/11/98*

*A B C Building
Construction Co.*

## REPORT OF NONCONFORMING WORK

Project :                            Report ref. No.:

Contract No. :                   Date of report :

---

Date nonconformity detected :

Location of nonconformity :

Nature of nonconformity :

Contract term / specification not satisfied :

Subcontractor(s) involved :

Disposition of nonconforming work :

Confirmed / probable cause of nonconformity :

Corrective action taken or to be taken, if any :

Other information :

Report prepared by :

............................
*Project Manager*

Date


Note : This report goes to Quality Assurance Manager.

---

*Form QF13-5*
*Issue 1, 27/11/98*

*Page 1 of 1*

*ABC Building*
*Construction Co.*

## QP 13.3 PROCEDURE FOR HANDLING OF CLIENT COMPLAINTS

**1 Purpose**
To define responsibility for review, and authority for settlement, of client complaints

**2 Scope**
Applicable to client complaints pertaining to construction in progress or execution of a contract

**3 Person responsible**
Project Manager

**4 Procedure**

4.1 The Project Manager reviews the nature of the complaint with the personnel concerned, including the subcontractor(s) involved.

4.2 Based on the outcome of the review, the Project Manager replies to the client or his representative, indicating the method of settlement.

4.3 The Project Manager takes corrective action in accordance with Quality Procedure QP14.1 if he considers it necessary.

4.4 The Project Manager records the complaint in the Record of Client Complaints (Form QF13-6) and signs off the entry after it is settled.

4.5 Except for minor complaints, the Project Manager prepares a Report of Client Complaint on Form QF13-7 and submits it to the Quality Assurance Manager.

**5 Records**
Form QF13-6   Record of client complaints
Form QF13-7   Report of client complaint

Procedure approved by   *A. T. Best*
                        ........................
                        QA Manager
Dated  27/11/98

*A B C Building*
*Construction Co.*

## RECORD OF CLIENT COMPLAINTS

Project :                               Client :

Contract No. :                  Architect / Engineer :

| Ref. No. | Date received | Verbal / Written | Nature of complaint | Settlement | Signature and date |
|---|---|---|---|---|---|
|  |  |  |  |  |  |
|  |  |  |  |  |  |
|  |  |  |  |  |  |
|  |  |  |  |  |  |
|  |  |  |  |  |  |
|  |  |  |  |  |  |
|  |  |  |  |  |  |
|  |  |  |  |  |  |
|  |  |  |  |  |  |
|  |  |  |  |  |  |
|  |  |  |  |  |  |
|  |  |  |  |  |  |

*Form QF13-6*
*Issue 1, 27/11/98*

# ABC Building Construction Co.

## REPORT OF CLIENT COMPLAINT

Project :                          Report ref. No.:

Contract No. :                     Date of report :

---

Complaint made by :

Date received :

Nature of complaint :

Contract term / specification not satisfied :

Subcontractor(s) involved :

Method of settlement :

Confirmed / probable cause of complaint :

Corrective action taken or to be taken, if any :

Other information :

Report prepared by :

............................
*Project Manager*

Date

Note : This report goes to Quality Assurance Manager.

*Form QF13-7*
*Issue 1, 27/11/98*

*A B C Building*
*Construction Co.*

## QP 14.1 PROCEDURE FOR CORRECTIVE ACTION

**1 Purpose**
To eliminate the causes of nonconformities including substantive client complaints

**2 Scope**
Applicable to nonconformities relating to product, process and quality system

**3 Person responsible**
Project Manager / Quality Assurance Manager

**4 Procedure**

4.1 On detection of nonconforming work or receipt of a client complaint, the Project Manager, assisted by the Engineer and the Quality Assurance Officer, investigates the situation that has led to the incident, identifying or confirming the cause of the nonconformity. The investigation normally looks into the following:
- provision of material, equipment and manpower for the process
- quality procedure(s) and work instruction(s) used and any deviations therefrom
- records of inter-communication between sections of the Company, subcontractors and the client's representative

4.2 The Project Manager takes appropriate action to rectify any inadequate or inappropriate provision for the process, including more stringent supervison if necessary, to ensure that the quality procedures and work instructions are strictly followed.

4.3 The Project Manager identifies any training need and either arranges for on-site training or requests the appropriate section to provide the training.

4.4 The Project Manager identifies any deficiency in the work instruction(s) issued under his authority and makes appropriate changes. Revised documents are subject to document control as per Quality Procedure QP5.1.

*ABC Building
Construction Co.*

4.5 The Project Manager records the results of the investigation and the corrective/preventive action taken in the Report of Nonconforming Work (Form QF13-5) or the Report of Client Complaints (Form QF13-7), and submits the completed form to the Quality Assurance Manager.

4.6 If any deficiency in the quality procedure(s) involved is identified, the Project Manager requests the QA Manager to make the necessary amendment. The request is made on Form QF14-1.

4.7 The QA Manager files the Report with other reports received previously from the same project and/or other projects. He enters the reported incident in the List of Nonconformance Reports (Form QF14-2) which forms the first page of the file.

4.8 The QA Manager scans through the List to determine whether the reported incident is recurrence of a nonconformity in the same project or another project. He evaluates the effectiveness of the corrective action in the light of similar action previously taken and advises the Project Manager accordingly if necessary.

4.9 If the reported incident is likely to occur in other projects, the QA Manager notifies the other project managers to take preventive action.

4.10 Should the nonconformity occur in the head-office operations, e.g. a mistake in a purchase order or incorrect distribution of a controlled document, the section head responsible for the process acts in the place of 'Project Manager' in the procedure described above.

4.11 The QA Manager summarizes all nonconformities discovered and correction actions taken and submits the information for the next management review.

**5 Records**
Form QF14-1   Corrective / preventive action request
Form QF14-2   List of nonconformance reports

Procedure approved by ......*A. T. Best*........
                                    QA Manager
Dated  27/11/98

*A B C Building
Construction Co.*

## CORRECTIVE / PREVENTIVE ACTION REQUEST

Request to :                                Request from :

Section / Site :                            Section / Site :

Description of problem :


Confirmed / probable cause :


Requested action :


Signature : ..............................        Date :

---

Requested action taken. *

Requested action not taken for reason(s) as follows. *

Alternative action taken as follows. *


Signature : ..............................        Date :


\*   *Delete whichever is inapplicable.*


*Form QF14-1*                                *Page 1 of 1*
*Issue 1, 27/11/98*

## ABC Building Construction Co.

## LIST OF NONCONFORMANCE REPORTS

| Report ref. No. | Nonconformity / Client complaint | Date of occurrence | Place of occurrence | Disposition | Corrective action |
|---|---|---|---|---|---|
| | | | | | |
| | | | | | |
| | | | | | |
| | | | | | |
| | | | | | |
| | | | | | |
| | | | | | |
| | | | | | |
| | | | | | |
| | | | | | |
| | | | | | |
| | | | | | |
| | | | | | |

*Form QF14-2*
*Issue 1, 27/11/98*

*A B C Building*
*Construction Co.*

QP 14.2   PROCEDURE FOR PREVENTIVE ACTION

**1   Purpose**
To eliminate the causes of potential nonconformities, including probable client complaints

**2   Scope**
Applicable to potential nonconformities relating to product, process and quality system

**3   Person responsible**
Project Manager / Section Head / Quality Assurance Manager

**4   Procedure**

4.1   A member of staff, either on site or in the head-office and irrespective of rank, who encounters a major / persistent problem or difficulty which may lead to potential nonconformity or client complaint can initiate preventive action by submitting a Corrective / Preventive Action Request on Form QF14-1. The request is directed to the Project Manager or Section Head under whom the staff member works, except for request for amendment of a quality procedure which goes to the Quality Assurance Manager.

4.2   The receiver of Form QF14-1 investigates the situation reported, identifying or confirming the cause of the problem or difficulty.

4.3   The receiver considers the action requested and decides either
(a)   to implement the action requested
(b)   to refuse the action requested, with reasons
(c)   to take alternative action.

4.4   The receiver endorses the form and forwards it to the Quality Assurance Manager for record. A copy of the form is returned to the staff member for information.

4.5   The QA Manager evaluates the effectiveness of the preventive action in the light of similar action previously taken and advises the Project Manager accordingly if necessary.

4.6   The QA Manager summarizes all preventive actions taken and submits the information for the next management review.

# ABC Building Construction Co.

**5** **Records**
   Form QF14-1   Corrective / preventive action request

Procedure approved by   *A. T. Best*
                        .........................
                        QA Manager

Dated   27/11/98

# ABC Building Construction Co.

## CORRECTIVE / PREVENTIVE ACTION REQUEST

Request to :                                Request from :

Section / Site :                            Section / Site :

Description of problem :

Confirmed / probable cause :

Requested action :

Signature : ...........................                Date :

---

Requested action taken. *

Requested action not taken for reason(s) as follows. *

Alternative action taken as follows. *

Signature : ...........................                Date :

\* *Delete whichever is inapplicable.*

*Form QF14-1*
*Issue 1, 27/11/98*

*A B C Building*
*Construction Co.*

## QP 15 PROCEDURE FOR HANDLING, STORAGE AND DELIVERY

**1 Purpose**
To prevent damage or deterioration of materials and semi-finished products during handling, storage and delivery

**2 Scope**
Applicable to purchased materials, including mechanical or electrical appliances, and precast / prefabricated units produced on-site / off-site, such as precast concrete elements, steel sub-assemblies and timber roof trusses

**3 Person responsible**
General Foreman

**4 Procedure**

4.1 The General Foreman establishes and maintains an inventory of each item of material or semi-finished product received on site, using Form QF-15, except for minor items in small quantities.

4.2 On receipt of a consignment of an item, the General Foreman arranges for receiving inspection as per Quality Procedure QP10.1.

4.3 The General Foreman, or a foreman acting on his behalf, directs the handling and delivery of the item to the designated storage area indicated on the site layout plan and, where applicable, ensures that the item is lifted at the specified points and transported by suitable means.

4.4 The General Foreman (or a foreman) enters the quantity received in the inventory of the item. By signing off the entry, he authorizes receipt of the item to storage.

4.5 The item received is stored either indoors or outdoors on level hard ground or a raised platform and in such a way as to comply with site safety regulations. Any special storage requirements indicated in the project quality plan must be satisfied.

4.6 When the item is required for use or installation, the General Foreman (or a foreman) enters the quantity removed in the inventory. By signing off the entry, he authorizes dispatch of the item from storage.

*A B C Building*
*Construction Co.*

4.7 At monthly intervals, the General Foreman inspects all items in storage or instructs a foreman to do so. On completion of stock-taking of an item, he remarks and countersigns (with date) against the last entry in the inventory. Any material that has deteriorated or been damaged beyond the acceptable quality is identified and reported to the Project Manager for disposition. Any discrepancy between the actual quantity in stock and the quantity on record is reported to the Project Manager who will investigate the cause and take corrective action if necessary.

5 **Records**
   Form QF15   Inventory Control of ..............

*A. T. Best*
Procedure approved by  ..........................
                        *QA Manager*
Dated  27/11/98

## ABC Building Construction Co.

### INVENTORY CONTROL OF ...................

Project :                              Client :

Contract No. :              Architect / Engineer :

| Date | Quantity in | Quantity out | Balance in stock | Authorized signature | Stock-taking (sign & date) |
|------|-------------|--------------|------------------|----------------------|----------------------------|
|      |             |              |                  |                      |                            |
|      |             |              |                  |                      |                            |
|      |             |              |                  |                      |                            |
|      |             |              |                  |                      |                            |
|      |             |              |                  |                      |                            |
|      |             |              |                  |                      |                            |
|      |             |              |                  |                      |                            |
|      |             |              |                  |                      |                            |
|      |             |              |                  |                      |                            |
|      |             |              |                  |                      |                            |
|      |             |              |                  |                      |                            |
|      |             |              |                  |                      |                            |

*Form QF15*
*Issue 1, 27/11/98*

*ABC Building*
*Construction Co.*

## QP 16 PROCEDURE FOR CONTROL OF QUALITY RECORDS

**1  Purpose**
To specify methods of identification, collection, indexing, filing, access, storage, maintenance and disposition of quality records

**2  Scope**
Applicable to quality records generated in the course of execution of construction contracts and operation of the quality system

**3  Person responsible**
Project Manager / Contracts Manager / Other Section heads

**4  Procedure**

4.1  With project-specific quality records, the Project Manager ensures that each document bears the contract number, the project title and date.

4.2  The quality records are kept in files as they are generated. Pertinent subcontractor quality records are similarly organized.

4.3  All record files are stored under lock and key and are accessible only with the permission of the Project Manager or his designate.

4.4  After completion of the contract, the Project Manager hands over all record files to the Contracts Manager for custody.

4.5  The Contracts Manager specifies a period of retention of each record file, normally seven years, and indicates it on the front cover of the file with the words 'This file shall be retained until ...............'

4.6  After expiry of the specified period, the Contracts Manager reviews the documents in the files and disposes of those that have no further value.

4.7  With quality records which are not related to any specific project, the head of the section where the records are generated ensures that each document bears a unique identification showing the subject matter and date, e.g. 'Minutes of Management Review dated ...............' and 'List of Acceptable Subcontractors for Concrete Work'. These records are filed in a similar manner. An index is established and maintained, indicating for each document the date after which it can be disposed of.

*Quality Procedure QP16*
*Issue 1, 27/11/98*

**$\mathcal{A}\mathcal{B}C$ Building
Construction Co.**

**5    Records**
    Index of quality records (non project-specific)

                             *$\mathcal{A}$. $\mathcal{T}$. Best*
Procedure approved by  ..........................
                             *QA Manager*
Dated   27/11/98

---

*Quality Procedure QP16*  
*Issue 1, 27/11/98*

*ABC Building*
*Construction Co.*

QP 17  PROCEDURE FOR INTERNAL QUALITY AUDITS

**1   Purpose**
To verify whether quality activities and related results comply with documented procedures and/or quality plan

**2   Scope**
Applicable to internal quality audits of both office operations and site operations

**3   Person responsible**
Quality Assurance Manager

**4   Procedure**

4.1  At the beginning of the calendar year, the Quality Assurance Manager drafts a tentative programme of internal quality audits for the whole year such that every section of the head office and every construction site is audited at least once during the year. In doing so, he takes into consideration the results of previous audits, especially the non-conformities noted. He also refers to the quality plans of individual projects for the frequency of quality auditing required. The audit programme is subject to change as need arises.

4.2  At least one month before a scheduled audit, the QA Manager selects one or more auditors from the current list of internal quality auditors. The auditors should be independent of those having direct responsibility for the activity being audited. If an audit team is formed, one member of the team is designated as the lead auditor.

4.3  The audit team, or single auditor as the case may be, studies the documented procedures and/or quality plan involved, and clarifies any doubtful points with the QA Manager. Checklists are prepared showing typically the following:
- persons to interview
- questions to ask of each person
- documents to check
- quality records to examine
- in case of site audit, areas of site and materials to inspect

Quality Procedure QP17
Issue 1, 27/11/98

*ABC Building
Construction Co.*

4.4. The audit team works out a time schedule of the audit, using Form QF17-1, which shows the activities to be evaluated, the persons to be present and the time allocated. The schedule is delivered, at least one week before the event, to the person in charge of the section or site to be audited, who may suggest changes to the schedule.

4.5 The audit team starts the audit by reviewing any nonconformities noted in the previous audit, the correction actions taken and their effectiveness.

4.6 Through interview, observation and inspection of records, the individual auditors seek evidence to confirm that the documented procedures and/or quality plan are strictly followed or otherwise. Any nonconformities noted are recorded. The auditors then jointly sort out their findings.

4.7 At the close of the audit, the lead auditor presents a verbal summary to key members of the section or site audited and invites them to respond. For any major nonconformity noted, the auditee is requested to suggest a corrective action.

4.8 The audit team prepares the audit report, using Form QF17-2, normally before leaving the location of audit. Page 1 of the form is completed in every case; page 2 is required when nonconformities are noted in the audit; page 3 is used for a follow-up audit if necessary. Each page of the form is signed by the lead auditor and acknowledged by the auditee. If corrective action cannot be formulated on the spot, a time limit is set for the auditee to make a proposal. Whether a follow-up audit is necessary is at the discretion of the lead auditor, taking into account the seriousness of the nonconformities and the corrective action decided upon. The audit report is produced in duplicate: one copy is given to the auditee and the other copy retained by the lead auditor.

*Note : If amendment of a quality procedure is considered necessary, a corrective action request is made to the QA Manager on Form QF14-1.*

4.9 Should a follow-up audit be required, the lead auditor arranges in due course for a member of the audit team to carry out the task. In the follow-up audit, the auditor verifies that the corrective action has been implemented and the nonconformity does not recur. The audit report is then completed by filling in page 3 of Form QF17-2.

*A B C Building*
*Construction Co.*

4.10 The lead auditor submits the completed audit report to the QA Manager.

4.11 The QA Manager examines the audit report and makes appropriate amendment to any quality procedure which is found to be impractical or inefficient.

**5 Records**
    Annual programme of internal quality audits
    Form QF17-1   Schedule of internal quality audit
    Form QF17-2   Report of internal quality audit

Procedure approved by   *A. T. Best*
                            *QA Manager*

Dated 27/11/98

*A B C Building
Construction Co.*

## SCHEDULE OF INTERNAL QUALITY AUDIT

Audit No. :

Audit date :

Audit location :

Auditor(s) :

Auditee :
*(Section head / Project manager)*

| Time | Activity | Staff to be present |
|------|----------|---------------------|
|      |          |                     |
|      |          |                     |
|      |          |                     |
|      |          |                     |
|      |          |                     |
|      |          |                     |
|      |          |                     |
|      |          |                     |

Prepared by  ..........................   Date  .................
                           *Auditor*

Agreed by  ..........................   Date  .................
                         *Auditee*

*Form QF17-1*
*Issue 1, 27/11/98*

# A B C Building Construction Co.

## REPORT OF INTERNAL QUALITY AUDIT

**General information**

| | |
|---|---|
| Audit No. : | Auditor(s) : |
| Audit date : | Auditee : |
| Audit location : | Previous audit No. & date : |

**Observations**

| Procedures audited | Comments |
|---|---|
|  |  |

Nonconformities noted ?      [ ] Yes      [ ] No

Reported by ............................. Date ..................
*Auditor*

Acknowledged by ............................. Date ..................
*Auditee*

*If nonconformities are noted, continue on page 2.*

*Form QF17-2*
*Issue 1, 27/11/98*

Page 1 of 3

# ABC Building Construction Co.

## Nonconformities

| Nonconformities | Cause, if known |
|---|---|
| | |

## Corrective action request

| Corrective action | Person responsible for action | Agreed date for action |
|---|---|---|
| | | |

Follow-up audit required ?     [ ] Yes     [ ] No

If yes, provisional date of follow-up audit  ..............................

Reported by  ............................. Date ...................
         *Auditor*

Acknowledged by ............................. Date ...................
         *Auditee*

*If follow-up audit is required, continue on page 3.*

*Form QF17-2*
*Issue 1, 27/11/98*

## *A B C Building Construction Co.*

### Follow-up audit

Date of audit :

Auditor :

| Corrective action | Evidence of implementation | Effectiveness |
|---|---|---|
|  |  |  |

### Further comments

..................................................................................................

..................................................................................................

..................................................................................................

..................................................................................................

Reported by                        Date
               ..........................            ..................
                    *Auditor*

Acknowledged by              Date
               ..........................            ..................
                    *Auditee*

*Form QF17-2*
*Issue 1, 27/11/98*

*A B C Building
Construction Co.*

**QP 18.1  PROCEDURE FOR TRAINING IN QUALITY SYSTEM**

**1  Purpose**
To familiarize the staff with the Company's quality system

**2  Scope**
Applicable to staff at all levels

**3  Person responsible**
Quality Assurance Manager

**4  Procedure**

4.1  The Quality Assurance Manager organizes and conducts in-house seminars on the Company's quality system at approximately six months intervals depending on demand.

4.2  Based on the participants expected, the QA Manager fixes the contents of the seminar, which typically include the quality assurance concepts, the Company's quality policy and the quality procedures related to the work of the participants. He may invite senior staff in the relevant sections to assist in the presentations.

4.3  The QA Manager publicizes the seminar within the Company at least three weeks before the event. Notice of the seminar is posted in the head office and at each construction site.

4.4  The Executive Officer arranges for all new staff, irrespective of rank, to attend the seminar as soon as practicable after taking up duty. Evidence of attendance is recorded on the employee's personal file.

4.5  All staff are encouraged to attend the seminar every two years. The QA Manager keeps a list of participants of each seminar and reminds them in due course to attend the seminar again. Should the action of a member of staff repeatedly result in nonconforming work or client complaint, he/she is required to attend the seminar when it is next held.

**5  Records**
List of participants of quality training seminar

Procedure approved by  *A. T. Best*
                                            QA Manager
Dated  27/11/98

Quality Procedure QP18.1
Issue 1, 27/11/98

*ABC Building
Construction Co.*

## QP 18.2 PROCEDURE FOR TRAINING IN QUALITY AUDITING

**1 Purpose**
To train internal quality auditors

**2 Scope**
Applicable to staff at all levels

**3 Person responsible**
Quality Assurance Manager

**4 Procedure**

4.1 The Quality Assurance Manager establishes and maintains a list of internal quality auditors which is made up of 10–15% of the Company's establishment.

4.2 The QA Manager selects representatives among the staff in various sections and at various levels to serve occasionally as internal quality auditors.

4.2 The QA Manager arranges for the selected staff to attend in turn a short course on internal quality auditing offered by a reputable organization.

4.3 The QA Manager organizes mock-up quality audits to further train the internal quality auditors if he deems it necessary.

4.4 The QA Manager keeps a list of internal quality auditors, showing their job titles, the sections they belong to and the dates they are enlisted.

4.5 The QA Manager removes an individual auditor from the current list who has served for three years in this capacity or has been found inefficient or dishonest in auditing. He then selects and trains another person to fill the vacancy.

**5 Records**
List of internal quality auditors

Procedure approved by ......*A. T. Best*......
                          *QA Manager*

Dated 27/11/98

*Quality Procedure QP18.2*
*Issue 1, 27/11/98*

*ABC Building*
*Construction Co.*

## QP 18.3 PROCEDURE FOR TRAINING IN OPERATIONAL / TECHNICAL SKILLS

**1 Purpose**
To identify training needs of staff in operational / technical skills and to provide appropriate training

**2 Scope**
Applicable to staff at all levels

**3 Person responsible**
Executive Officer

**4 Procedure**

4.1 The Executive Officer establishes and maintains a training record, using Form QF18-1, for each member of staff irrespective of rank. The training record is kept in the personal file of the staff member and shows his/her qualification, experience and training compared to the job requirements. Any deficiency in skills is made known to the immediate supervisor.

4.2 The supervisor, with the assistance of the Executive Officer if necessary, arranges for the necessary training as soon as practicable after appointment of the staff member and before he/she is to work independently. The training provided may be an external training course, an in-house workshop or on-the-job training, as appropriate.

4.3 After receiving the training, the trainee submits a training report, using Form QF18-2, to the Executive Officer through the supervisor. Satisfactory completion of the training is entered into his/her training record.

4.4 Before assigning someone to a task which requires special skill, the supervisor informs the Executive Officer of the training needs and they jointly work out a training programme for the assignee to acquire the skill in time for the task.

4.5 On or before transfer / promotion of a staff member, the new supervisor, in collaboration with the Executive Officer if appropriate, arranges for the necessary training, as above, to cover any deficiency in skills required by the new position.

*A B C Building*
*Construction Co.*

**5   Records**
   Form QF18-1   Training Record
   Form QF18-2   Training Report

                            *A. T. Best*
Procedure approved by  ........................
                            *QA Manager*

Dated  27/11/98

---

*Quality Procedure QP18.3*
*Issue 1, 27/11/98*

# *A B C Building Construction Co.*

## TRAINING RECORD

### Personal details

| | |
|---|---|
| Name : | Job Title : |
| Staff No. : | Section / Site : |
| Date appointed : | Supervisor : |

### Job requirements

Qualification :

Skills / Experience :

### Pre-appointment status

Qualification :

Skills / Experience :

### Training required

In-house :

External :

### Training provided *(Continue on new page if required.)*

Course / Programme :

Organizer / Trainer :

Date completed :

*Form QF18-1*
*Issue 1, 27/11/98*

## ABC Building Construction Co.

## TRAINING REPORT

**Particulars of trainee**

Name :

Job Title :

Section / Site :

Reason for taking up training :
*(Job requirement, special duty, upgrading, etc.)*

**Particulars of training**

Training course / Programme :

Organizer / Trainer :
*(Indicate trainer or section if in-house.)*

Main contents :

Duration :

Certificate gained, if any :

**Endorsement**

I have completed the above-mentioned training and found it

[ ] very useful  [ ] useful  [ ] not useful

Signed : ..............................   Date : ..................
    *Trainee*

---

My observation indicates that the performance of the trainee after the training is

[ ] much improved  [ ] improved  [ ] not improved

Signed : ..............................   Date : ..................
    *Supervisor*

*Form QF18-2*
*Issue 1, 27/11/98*

# References

Ashford, J.L. (1989) *The management of quality in construction,* E & FN Spon, London.

Baldwin, A.N., Thorpe, A., and Alkabi, J.A. (1994) Improved materials management through bar-coding: results and implications from a feasibility study, *Proceedings, Institution of Civil Engineers,* **102**, (4), pp. 156–62.

Barber, J.N. (1992) *Quality management in construction - contractual aspects,* Special Publication 84, Construction Industry Research and Information Association, London.

BRE (1982) *Quality in traditional housing,* **1–3**, Building Research Establishment, HMSO, London.

BSI (1979) *BS5750: Quality systems, Parts 1, 2 and 3,* revised 1987, withdrawn 1994, British Standards Institution, London.

CIOB (1987) *Quality assurance in building,* Chartered Institute of Building, London.

CIRIA (1989) *Quality assurance in construction,* Special Publication 64, Construction Industry Research and Information Association, London.

IQA (1995) *Quality systems in the small firm,* Institute of Quality Assurance, London.

ISO (1991) *ISO 10011: 1991 Guidelines for auditing quality systems – Parts 1, 2 and 3,* International Organization for Standardization.

ISO (1994a) *ISO 8402: 1994 Quality management and quality assurance – Vocabulary,* International Organization for Standardization.

ISO (1994b) *ISO 9004-1: 1994 Quality management and quality system elements – Part 1: Guidelines,* International Organization for Standardization.

ISO (1998) *ISO/CD1 9001:2000 Quality management systems – Requirements,* International Organization for Standardization.

Kettlewell, D.H. (1990) Quality assurance - it will improve things won't it? *Proceedings of Conference on Quality Assurance for the Chief Executive, Institution of Civil Engineers,* Thomas Telford, London, pp. 37–46.

Low, S.P. (1998) *ISO 9000 and the construction industry*, Chandos Publishing, Oxford.

Novack, J.L. (1994) *The ISO 9000 documentation toolkit*, P T R Prentice Hall, New Jersey.

Roberts, R.J.F. (1991) Quality does not cost – it pays, *Australian Construction Law Reporter*, **10** (4), pp. 137–44.

SA/SNZ (1996) *SAA/SNZ HB66:1996 ISO9000 Guide for small businesses*, Standards Australia / Standards New Zealand.

SA/SNZ (1997) *AS/NZS 3905.2: 1997: Quality system guidelines, Part 2: Guide to AS/NZS ISO9001, AS/NZS ISO9002, and AS/NZS ISO9003 for construction*, Standards Australia / Standards New Zealand, pp. 15–17 & p. 95.

Tam, C.M. (1996) Benefits and costs of the implementation of ISO 9000 in the construction industry of Hong Kong. *Journal of Real Estate and Construction*, (6), pp. 53–66.

# Index

**Bold** refers to figures.
*Italic* refers to tables.

Accreditation
  of auditor   104
  of certification body   86
Accreditation body   86
Action plan   60, **61**
Audit
  frequency   73–4, 95
  process   76–81
  programme   74, **75**
  report   79–81, 92–3
  schedule   76, **77**
Audit, follow-up audit   79, 93
Audit, internal   35, *41*, 71–2, **75**, **83**, 147, 234–40
Audit, project   73
Audit, surveillance   95–6
Audit, system   73, 84–5, 92–5
Audit, third party   82–3, 103–4
Auditor   72–3, 103–4
  attributes of   72–3
  training of   73, 242
Auditor, lead   74, 91

Benefits of quality assurance   8–9, 99
Building construction, *see* Construction
Business, small   105–7

Calibration   29
  *see also* Control of inspection, measuring and test equipment
Certificate   86, 93, 95
Certification   **61**, 85, 93
  process of   88–93
  scope of   87–8
  time and cost of   96–7

Certification, conditional   93
Certification body   59, 85–7
  accreditation   86
  selection   85–7
Checklist   76, 102, 168–9, 171–2, 174–5
Clerk-of-works   104–5
Client (customer) complaint   31, 145–6, 219–21
Communication   8, 99, 106
Conformance, *see* Nonconformance
Construction   3, 7, 17–18, 27–30, 42
Contract   5, 52–3
Contract review   20–1, *38*, 141, 170–2
Contractor   44, 53, **56**, 105
Control
  of customer-supplied product   27, *39*, 143, 191–2, 143, 191
  of document and data   21–5, *38*, 141–2, 176–82
  of inspection, measuring and test equipment   30–1, *40*, 145, 206–8
  of nonconforming product (work, supply)   32, *40*, 145–6, 210–18
  of quality records   34, *41*, 147, 232–3
Controlled copy   22, **24**
Controlled document   21–2, **23**
Corrective action   33, *41*, 146, 222–5
Corrective action request   79–**80**, 224

Cost of quality assurance 8–9
Cost of third party certification 97

Delivery, *see* Handling
Design control 21, *38*, 141
Disposition of nonconformity 32–3, *40*, 145–6, 210–18
Distribution of document, *see* Document distribution
Document control 21–5, *38*, 141–2, 176–82
Document distribution 22, *38*, 141–2, 176–82

Evaluation
 of quality system, *see* Audit
 of subcontractor 26, 142–3, 183–6
 of supplier 26, 142–3, 187–90

Forms 69, 102, 120

Handling 34, *41*, 146, 229–31
Hold point 29–30, 50, **51**

Inspection and test plan 29–30, 44, **46**, 49–50, **51**, 166
Inspection and test status 31, *40*, 145, 209
Inspection and testing
 final 30, *40*, 144, 203–5
 in-process 29, *40*, 144, 200–2
 receiving 29, *40*, 144, 197–9
Internal quality audit, *see* Audit
International Organization for Standardization 14
ISO, *see* International Organization for Standardization
ISO 9000
 family of standards 15–18, 58, 100, 105, 107–8
 related standards *17*

ISO 14000 107
ITP, *see* Inspection and test plan

Juran, J. M. 3

Management
 representative 19
 responsibility 18, *38*
 review 19, 81–2, 139, 154
Manual, *see* Quality manual
Master list 20, 22, **23**
Motivation 68
Motivator 68
Nonconformance 78–9, 95, 102
Nonconformance notice 33, 212, 216
Nonconforming product (work, supply) 32, *40*, 145–6, 210–18
Nonconformity 32–3, 78–9
 *see also* Disposition of nonconformity

Organization chart 18, 116, **132**
Organization structure 11, 131

Preventive action 33, *41*, 146, 226–8
Procedure, *see* Quality procedure
Process control 26–27, *39*, 143–4, 194–6
Product, customer-supplied 27, *39*, 143, 191–2
Product identification and traceability 25, *37*, 143, 193
Project quality management, *see* Quality management
Purchaser 25, 53
Purchaser-supplier interaction 53–4, **56**
Purchasing 25, *39*

Quality
 definition 3

Index   251

of building construction   3–4
Quality assurance   5–7, 52, 57, 109
Quality audit, *see* Audit
Quality consultant   59
Quality control   4–5, 109
Quality function   *38*–41, 117, 139–48
Quality management   11, 42, 107, 117
Quality management review, *see* Management review
Quality manager   19, **136**
Quality manual   14, 19, 63–4, 129
Quality plan, *see* Quality planning
Quality planning   20, 42, 45–8, 149–60, 158–66
Quality policy   13, 18, 114–16, 130
Quality procedure   13, 19, 64, 118–20, 122–3
Quality records   34, *41*, 48, 94, 147, 232–3
Quality system   7, 11, **12**, 19, *38*, 139
  development   58, **61**
  documents   **13**, *55*, 63–5, 113, 122
  implementation   67
  requirements   18, *38–41*, 63
Quality system standard   14
  *see also* ISO 9000

Records, *see specific records*
Registrar, *see* Certification body
Registration, *see* Certification
Review, *see specific review*
Revision of ISO 9000   17, 107–8

Servicing   36, *41*, 148
Small business, 105–7
Special process   28
Statistical techniques   36–7, *41*, 148
Storage, *see* Handling
Subcontract   26
Subcontractor   25, 43–4, 105
  accceptable   26, 142–3, 183–6
  evaluation   26, 142–3, 183–6
Supervision   5, 28, 52
  *see also* Process control
Supervisor   44, 52
Supplier   14, 25, 43, 53
  acceptable   26, 142–3, 187–90
  evaluation   26, 142–3, 187–90
SWOT analysis   99

Temporary works   21, 28, 141
Tender review   20, *38*, 141
Total quality management   108, **109**
TQM, *see* Total quality management
Training   36, *41*, 65, 70, 147, 243–6
  of quality auditor   65, 73, 242
Training programme   36
Training records   36, 245

Variation review   21, *38*, 141, 173–5
Verification   11, 26, 50, 165
  *see also* Inspection and testing

Witness point   30, 50, **51**
Work instruction   13, 19, 64

8794

TH
4382
.C48

1999